Emergent Directions in Engineering Education

Volume 2

The book series *Emergent Directions in Engineering Education* provides a dedicated forum to explore innovative and transformative developments, highlighting emerging visions and future-oriented directions, in engineering education. Driven by advances in digital technologies, interdisciplinary practices, and increasing attention to social, ethical, and sustainability considerations, engineering education is evolving rapidly. This series captures and synthesises these dynamic changes through scholarly contributions that bridge research, theory, and practice.

The series invites conceptual and philosophical perspectives, rigorous theoretical discussions, research syntheses, and empirical studies, across all themes of relevance to engineering education, which aim to foster a deeper understanding and advancing of the field. By presenting scholarship in an accessible and integrated manner, this series encourages dialogue among engineering educators, researchers, and practitioners across diverse sub-disciplines.

Designed for engineering education researchers at all career stages, as well as faculty, instructors, curriculum designers, academic administrators, industry professionals, and policymakers, the series uniquely addresses a critical gap by providing a platform for comprehensive exploration and interdisciplinary engagement. It encourages submissions from a diverse authorship, ensuring openness to new ideas and perspectives from across the global engineering education community. Book proposals for this series may be submitted to the Publishing Editor: Claudia Acuna E-mail: Claudia.Acuna@Springer.com

Anette Kolmos

Engineering Education for Interdisciplinary Futures

Problem- and Project-Based Learning for a Chaotic World

Springer

Anette Kolmos
Aalborg University
Aalborg, Denmark

ISSN 3091-3136 ISSN 3091-3144 (electronic)
Emergent Directions in Engineering Education
ISBN 978-3-032-32539-6 ISBN 978-3-032-32540-2 (eBook)
https://doi.org/10.1007/978-3-032-32540-2

This work was supported by The Obel Family Foundation and the Poul Due Jensen Foundation/Grundfos Foundation.

This Springer imprint is published by the registered company Springer Nature Switzerland AG
The registered company address is: Gewerbestrasse 11, 6330 Cham, Switzerland

Foreword

This is a book about how we might prepare engineering students for chaos. Chaos is front-page news now, with major conflicts in the Middle East and Ukraine causing widespread disruption to millions of lives and to the global economy. However, these conflicts distract us from the ongoing chaos generated by climate change – floods, fires, famines, pandemics, all occurring more frequently than in the past and with greater magnitude.

What is chaos in this context? Chaos can be crisis. Think of thousands of people displaced by war, floods, earthquakes, fires, etc. How do we respond as societies to support our fellow citizens and repair broken infrastructure that can lead to death and disease? There is important work here for engineers in many corners of the globe – roads, buildings, airports, power systems, telecommunication systems, wastewater systems, all the work of engineers of every discipline.

However, apart from these crisis situations, there are also unpredictable aspects of complex engineered systems that can lead to chaotic situations, where the systems we rely on rapidly break down. A sudden equipment failure in a power system can quickly lead to other generators dropping offline, leading to a grid-wide power outage. Similar problems occur in telecommunications systems, the Internet, airline booking systems, and many more.

These systems are increasingly being hardened to resist such outages. Nevertheless, as systems become more complex, made up of millions of individual devices and tens of millions of lines of code, it becomes increasingly difficult to predict the system behaviour. Inside this complexity lies potential chaos and disruption. The work of engineers is to minimise this possibility. Engineering education must take this into account.

Chaos is the golden thread through this book. How might we prepare our engineering students to design and maintain these increasingly complex engineered systems?

However, this book is much more than chaos. The book is also a history of problem-based learning, told from the perspective of the research and developments at Aalborg University in Denmark, which have inspired many of us around the world to implement similar ideas at our own universities. Anette Kolmos has been at the centre of those developments through her establishment and leadership of the UNESCO Centre for Problem-Based Learning from 2007, initially as UNESCO Chair.

In this book, Anette leads us through the origins of PBL as a quest for a more democratic form of education in the 1970s, through its adoption across a range of professional disciplines, where it became centred on holistically developing professional competencies in realistic situations (the problems and projects).

As the decades have passed, we have realised the need to challenge students with increasingly complex problems. This trend has been assisted by increasingly complex software systems that have simplified information searches and provided students with increasingly sophisticated simulation tools, allowing them to develop solutions in a fraction of the time previously required. Manufacturing technologies such as 3D printing also allow students to bring their ideas to reality in record time. Artificial intelligence is only accelerating these trends.

This book, then, takes the reader through this historical transition from the social changes of the 1960s and 1970s to a new form of professional education intended to better prepare our graduates for the world around them. Complexity and chaos are the new challenges that our graduates will face. There is no one better qualified than Anette Kolmos to narrate this transition. Enjoy!

Melbourne, Australia
April, 2026

Roger Hadgraft

Preface

I have spent at least two years writing this book. There have been several versions, from a higher doctorate on problem- and project-based learning (PBL) to reviewing experiences in essays and finally deciding on the current version. I wanted to summarise my experiences and research from setting up a PBL university and the establishment of an entirely new university culture. What lessons have been learned? How can these PBL lessons be applied to current issues for students who are concerned about their future due to climate, societal, and technological chaos?

I've been studying problem- and project-based learning (PBL) in engineering education for over 40 years. I am curious about what we can learn from these change and transformation processes and if these experiences are relevant today. I find that it is important to draw lessons from previous and existing PBL practices, and there is no doubt that students' learning to identify problems in teams can be beneficial when dealing with chaotic issues, both in terms of identifying challenges and learning to collaborate together within teams.

The current PBL curricula, however, will not suffice in preparing students for a resilient response to chaos. In this book, I emphasise the importance of having problem-oriented scientific knowledge, risk management, interdisciplinary and transdisciplinary knowledge collaboration, and of learning boundary crossing, the ability to adapt to new situations, and collaborative and individual resilience.

There is a need for the next phase of engineering education to embrace and integrate both new cognitive and emotional aspects. Education must constantly adapt and address the needs and issues of contemporary society. No matter what we think of more traditionally taught curricula or versions of PBL curricula, there will be a need for reflection and development. Student-centred curricula that incorporate PBL or other active learning methodologies may have deviated from their intended path. There is always a risk of institutional stagnation and instrumentalism.

In this book I advocate for a new development, and I believe that the lessons to be learned from the establishment of PBL reform universities in the 1970s and the subsequent development of engineering education can be applied. I am calling for a new wave of collaborative pioneers who can transform education to the next stage, where students learn to live in an ever more complex and chaotic world. I hope that this book will spark a new conversation about the future of engineering education, which in my view is about educating the whole person.

I use Aalborg University as an example because this is where I obtained most of my research data and experience. For over 40 years, I have represented PBL as the chair and founding director of the Aalborg Centre for Problem Based Learning in Engineering Education and Sustainability, under the auspices of UNESCO. However, I have only been a visible external face. The development at Aalborg University has been carried out by an engaged and deeply motivated academic staff throughout all levels of the organisation and engineering departments. I wanted to mention names because I believe my colleagues deserve recognition and respect. Many of them have never written educational articles about what they have done, and their names will be forgotten despite their significant contributions. I realise that after a list of 50 names, this would be risky and unfair because I may have overlooked some important contributors along the way. As a result, I closed with an inclusive thank you to my colleagues at Aalborg University.

I would, however, like to name and thank colleagues from the InterPBL project (Interdisciplinary PBL methodologies in engineering education) funded by the Poul Due Jensen Foundation, Jette Egelund Holgaard, Henrik Worm Routhe, and adjunct professor Roger Hadgraft, UTS, for their encouragement and constructive feedback throughout the process. I'd like to thank Max Denyer-Green for being patient with my English and pointing out the unreadable sentences. Furthermore, I would like to thank my husband of 40 years, Peter Henningsen, for his love and support throughout both the difficult and energetic periods of life.

Finally, I thank the Poul Due Jensen Foundation and the Obel Family Foundation for their financial support in writing this book.

Aalborg, Denmark Anette Kolmos
March 2026

Competing Interests The author has no competing interests to declare that are relevant to the content of this manuscript.

Contents

About the Author

Anette Kolmos is a Strategic Advisor, Professor in Engineering Education and PBL (2003-2026), Aalborg University, Founding Director (Director 2014–2023) for the UNESCO category 2 Centre: Aalborg Centre for Problem Based Learning in Engineering Science and Sustainability, UNESCO Chair in PBL in EE (2006–2014), Guest professor at KTH, Stockholm (2012–2017), and University Technology Petronas (2026–2028). She was President of SEFI 2009–2011 (European Society for Engineering Education) and Founding Chair of the SEFI Working Group on Engineering Education Research. She was awarded the SEFI Leonardo da Vinci Medal in 2023, IFEES Global Award for Excellence in Engineering Education, 2013, and the SEFI fellowship in 2015. During the last 20 years, Dr. Kolmos has researched the following areas, primarily within Engineering Education: gender and technology, project-based and problem-based curriculum (PBL), generic competencies, educational change, and interdisciplinarity. She is currently PI for the research project InterPBL on interdisciplinary projects (2020–2026) funded by the Grundfos Foundation.

Chapter 1
Motivation

1.1 Introduction

Many young individuals feel anxious about the future, a challenging emotion to articulate that could hinder their learning. Fear and other such emotions, albeit irrational, significantly influence individuals' motivation, values, and overall well-being. What started my thinking of new challenges to the educational system was a new television series by renowned Danish filmmaker Thomas Vinterberg that portrays a future in which Danish and Dutch individuals become climate refugees due to rising sea levels. These individuals forfeit all their assets, and some of the consequences include real estate becoming unsellable, all Danish stocks falling in value, and families being split among several countries. Suddenly, the Danish white middle-class populace experiences a decline in status and wealth, becoming reliant on less wealthy nations. This may seem overstated, but reality often surpasses the cinema, as evidenced during the Covid-19 pandemic. War in Europe, power cuts, terror, political instabilities—the list of possible disasters is suddenly much longer. Disasters have always been around, but the information level of today influences the individual's beliefs in the future.

This led me to read much more about how the educational system actually handles students' anxieties for possible chaotic futures. In this book, I will argue that it is the entire educational system's responsibility to address student anxiety about potential future chaos, and one potential solution is to rethink some of the student-centered learning methodologies that have been implemented throughout the past 50 years. Student-centered learning should in principle put the student in the center; however, this might not always happen although the intentions were in that direction. With this in mind we can do better in preparing students for chaos.

I have been working with problem- and project-based learning (PBL) as both an educator and a researcher for 41 years, which equating to a lifetime of experience. However, I also experienced the PBL reform as a student, starting in 1976, so I have

A. Kolmos, *Engineering Education for Interdisciplinary Futures*, Emergent Directions in Engineering Education 2,
https://doi.org/10.1007/978-3-032-32540-2_1

been involved in and experienced PBL practices for 50 years. PBL ideas and practices have changed during the period from the 1970s to the 2020s, and this is the story I would like to share in this book, together with a prospect of how PBL can contribute to the need for developing a contemporary pedagogy for the future.

During my career, I have built up the UNESCO Chair on Problem Based Learning (2005–2014) and the UNESCO Category 2 Centre: The Aalborg Centre for PBL in Engineering, Science and Sustainability (2014–2023). These two organizations have been affiliated with Aalborg University (AAU), and my primary research on PBL is based on the PBL practices at AAU. The primary global endeavor involved organizing the International Research Symposiums on PBL (IRSPBL), a series of events that have taken place on nearly every continent. I have conducted research on PBL, educating PhD students as change agents, and running training for institutions that wanted to upgrade or educate their academic staff. This has included a lot of travel.

My career was based on the early PBL pioneers, and I feel their ideas, contributions, and publications should be acknowledged as a foundation for the UNESCO Centre for PBL in Engineering, Science and Sustainability, as well as the achievements we have made both internally at AAU and globally. I might be the most cited for PBL at AAU, but I did not do the hard work in building up the system. The development of AAU-PBL is the result of a sincere commitment among my AAU colleagues. There has been a pioneering spirit saturating the culture, where different practices have been tested but not always written down. My contribution has been to describe the practices and research on the impact of these new learning models together with ideas for enhancement.

Participating in that pioneering culture during the 1980s and 1990s has been crucial for my professional career and the cultivation of professional confidence. A contributing component to the AAU-PBL was my belief in my research and practice beyond my office door. What I define as integrity is the relationship and interaction between research, practice, and personal values. It encompassed not merely statistics and theory, but also practical application and real-life experience.

The students' learning behaviors immediately outside the classroom corroborated the findings about the value of PBL in relation to alternative pedagogical approaches. As a researcher, I frequently engaged in casual discussions with the students, gaining substantial insights into the issues they were addressing and the solutions they presented. Their expressions and gaze conveyed more than research could articulate: a profound commitment and belief in their endeavors. There may be instances when they seem more discouraged and frustrated, making it challenging to elicit responses from them; if they do respond, it may pertain to feelings of confusion and lack of direction. There were instances where students within the teams were isolated from their peers. However, despite difficulties, they learn to establish direction through collaboration.

Learning is also about taking emotional ownership of the process of learning in addition to scientific and professional development. This intrinsic drive is developed at the individual level within a collaborative culture. This is not akin to being solitary in the classroom with a singular assignment destined solely for submission to the academic instructor; on the contrary, it cultivates a collaborative work identity and fosters a sense of collective belonging yet preserves individual identity.

Collaborative learning methods may encounter challenges; you might be misunderstood or misread, you may occasionally receive positive feedback, or you might feel that the response to your work is irrelevant or unproductive. As a member of a project team, you will receive feedback to facilitate the advancement of your thought processes. Perhaps we have not yet developed an understanding of learning that reconciles the differences between individual students and the collaborative learning culture throughout the years. The pragmatic development has concentrated on creating instruments for collaborative learning rather than on people's comprehension of their identities within teams.

I have contributed to building up a research community on engineering education and PBL nationally as well as internationally. The scope of the research might sometimes extend beyond engineering education, as numerous challenges associated with implementing PBL are relevant to various other disciplines. Engineering students have similar challenges in interdisciplinary collaboration to those of social science students, and cultural and scientific differences influence learning methods and cognitive comprehension. The way other disciplines are perceived will be influenced by things like how to discuss problems, how to organize the collaborative process, how much time to spend identifying the problem, how to approach problem-solving, when to meet in the morning, and how to make a report. There are fundamental disciplinary differences beyond scientific knowledge in terms of both culture and learning.

From the beginning, we had to argue for why we should have funding and positions to research engineering education. This had low prestige in the academic world of engineering and science. Internationally, PBL was something mysterious and not trustworthy. My pioneer colleagues and I always had to make a case for PBL and for us this was learning to stand up in strong headwinds. There is a saying that nobody can really understand the battlefield unless you have been there. We were working to gain respect and acknowledgement of both engineering education research (EER) and PBL. It has been fun, with the energy coming from the communities that we established in Denmark, at the Nordic level, at the European level, and globally. This has also been a personal investment beyond any paid job.

For one of my first activities as a young scholar, I was asked to evaluate the new engineering education practices at Aalborg University, which at the time lacked clear concepts. This track has persisted over the years, focusing on comprehending various project kinds and providing a conceptual framework to articulate the practice while also encouraging the advancement of interdisciplinary initiatives.

Compared to many other universities, the arguments for research on engineering education at AAU were easier, as the university needed documentation of the learning process. As a reform university established in 1974, it took time to gain the societal acknowledgement that, for example, engineers who graduated from AAU could become members of the Danish Association for Engineers. The Deans welcomed the documentation of how the companies received engineering graduates. Additionally, there was an interest in the students' experiences with collaborative learning. That did not mean there was a high number of positions. At AAU we had some academic positions and were able to participate in the growth of EER in the US during the years from 2000 to 2009. In the following years, European countries invested more in the development of EER. The EER and PBL research fields overlap and reach out to different communities. The core of my research and activities has been within engineering education.

1.2 The Narrative in the Book

In this book there are five main focal points:

1. Setting the scene, which describes the contemporary challenges.
2. Core elements in PBL that look into the history, theory, and added value in terms of learning generic competencies.
3. Emerging trends that embrace future developments such as interdisciplinary learning, boundary crossing, resilience, and emotions.
4. Management and transformation containing theories, strategies, and stories of existing change processes.
5. Concluding remarks which point at strategies for the further development of PBL and student-centered learning for inclusion of learning to live-with chaos.

1.2.1 Setting the Scene

This book's narrative begins in Chap. 2 with an attempt to understand both young people's anxiety and our research colleagues' concerns about the future. An emerging new trend about pedagogies of collapse, which is based on a much broader conceptual understanding of the role of university education in cognition and emotions, and knowledge and society. The Cynefin framework of simple, complicated, complex, and chaotic problems is used to outline the various types of chaotic and complex problems that educators face.

1.2.2 Core Elements in PBL

Chapter 3 covers the establishment of the reform universities during the 1970s and a comparison of very different PBL curricula models. It is inspirational to identify the variations in student-centered learning models, however, each model has become institutionalized with the risk of becoming monotone and static.

Chapter 4 is the example of Aalborg University with the experiences of contributing to the building up of the university and the methods that the Aalborg PBL model developed for problem analyses, problem solving, and assessment. The chapter finishes with an overview of the critiques that followed the reform universities, and which still exist.

Chapter 5 describes the PBL models based on sets of PBL learning principles and are reflected in the learning theories, which are useful for understanding complex learning settings. Most of the PBL practices are developed based on trial-and-error experiments in the classroom; however, behind the models there are learning theories that explain some of the learning processes. Formulating principles and reflecting them in learning theories provides an overview of collaborative learning processes at a more abstract level. This also opens up for a greater variation at the concrete curriculum level.

Chapter 6 outlines the acquisition of generic competencies. There is an introduction on the increasing number of generic competencies and skills in higher education, followed by a focus on the learning of generic competencies through reflection and experimentation. Variation is highlighted as a key to learning these types of practice competencies.

1.2.3 Emerging Trends

Chapter 7 concerns the interdisciplinarity of learning and collaboration, which will inevitably be part of future curricula. All problem-solving strategies will be interdisciplinary, and we will have to be able to act across the boundaries in chaotic situations. Questions like how we understand interdisciplinarity and transdisciplinarity will be addressed, and how can this be reflected in various types of student projects?

Chapter 8 draws attention to boundary crossing as one of the future competencies for engineers. Crossing boundaries is necessary in many situations, across disciplines, cultures, languages, community practices, etc. Boundary-crossing capabilities are needed in order to solve chaotic situations, and students should have an experience with this in education.

1.2.4 Management and Transformation

Chapter 9 addresses change and transformation in higher education combined with leadership and management strategies. Universities should provide both challenges and safe spots for students. This also counts for the development of the university culture, which ought to be exemplary for students. To integrate interdisciplinary learning, boundary crossing and the development of student resilience are components of a cultural shift at universities. This is much more than just educational change; it could be an educational transformation that represents new values.

Chapter 10 examines the power of networks as a staff development strategy for universities. There are numerous educational development units that do not have the greatest internal impact at their respective institutions. These become stronger in a network and have the potential to influence broader educational transformations.

1.2.5 Concluding Remarks

As a conclusion, Chap. 11 highlights the essential elements of engineering curricula. It is found that while PBL has many advantages, there are certain areas that require further development. Some of the topics discussed include problem-oriented science, risk management integration, interdisciplinary and transdisciplinary collaboration, crossing boundaries, and resilience.

1.3 Reflection 1: Living-With Chaos

After each chapter, I allow myself to reflect, my immediate thoughts and how it relates to contemporary challenges of chaos. What can we learn from history? What perspectives should we anticipate encountering? It is an attempt to keep the reader focused throughout the book on the need to be talking about living-with chaos. This is a relatively new topic for my generation to discuss. I was born in the 1950s, attended school in the 1960s and 1970s, and started university in 1976. My childhood and youth were spent between the aftermath of war and the fear of another war, as well as economic optimism and a strong emerging youth culture of interdependence.

When I sat at my family's dinner table I learned about the First and Second World Wars, and when I was with my friends, I felt an infinite amount of freedom and the liberty to make my own decisions. We were empowered by each other, by the new music, new ways of living and we believed that we could make a difference. We also grew up without the internet and social media. We listened to Radio Luxembourg, recorded songs on our old reel-to-reel tape recorders, and drank tea while talking about a free alternative world. At university, we learned to be quite critical of the society determined by the capitalist economy. We took control, were activists, and occupied the vice chancellor's office in our struggle for student rights. We had a great sense of agency and believed we were masters in our own lives.

The reality was significantly more complicated than this painted picture, but it is critical to understand today's young generations, who grew up with the internet and social media and can experience constant peer review and social pressure. Conversations on social media can be rude and direct, with no nuance. The information overload is immense. Every morning, we learn about the latest news from the world's top leaders, such as US pressure to take over Greenland, the war in Ukraine, the Palestine issue, a war with Iran, the treatment of a new virus discovered in Asia,

ICE killing US citizens, the inflation rate, rising food prices, the situation with NATO, activities at Finland's border with Russia, the flooding in the UK caused by rain, and so on. My feeling is that I become terrified because I can't do anything to help in these situations. The world seems chaotic for all generations.

For my generation, working with ill-structured problems were an advantage and we were motivated by pursuing our interests. How will it be for the young generations today? We have research that documents the positive effects of different aspects of PBL. However, we have never related PBL to learning to react to chaos.

Chapter 2
Chaos and Concerns for the Future

2.1 Anxiety over Climate Change

Anxiety is part of life and is one of the emotional poles. Nobody will get through life without having moments of anxiety as well as moments of joy and pleasure. Emotionally, there will always be ups and downs. Most people of various generations would have experienced the feeling of anxiety, for example, during one of the two World Wars or the Cold War, or in situations of terror when one would not know where the next attack would be. There have always been elements of uncertainty and unpredictable chaotic situations.

Social research has shed light on anxiety among young people who also have an increasing number of diagnoses such as ADHD, autism, etc. Around 11% of 11–17 years old children in the United States have been diagnosed with ADHD, many of whom also go to university and are involved in engineering education. My generation questions whether this anxiety was always there. Perhaps in this society of control, we test and diagnose more. Is it the pesticides in food and the environment? Did we bring up a generation of young people who are more scared of conflicts and possible disasters than we have ever been?

When I started to review the literature on anxiety among young people, much of what I read provided food for thought. The research literature reports on anxiety of climate, war, and other disasters. The majority of literature outlines the fear of climate change.

> It is completely unfair to ask a generation of young people to develop enhanced psychological stamina to face climate change. But because of decades of inaction on the part of previous generations, they are being asked to do so now. We owe it to children and young people to prioritize the mitigation of climate change at its source, while at the same time investing in evidence-based tools to support their mental wellbeing in the face of this ongoing crisis. In the process, we might see the generation of young people living in this era of unprecedented uncertainty become the most resilient generation yet. (Hurley et al., 2022, p. 190)

A. Kolmos, *Engineering Education for Interdisciplinary Futures*, Emergent Directions in Engineering Education 2,
https://doi.org/10.1007/978-3-032-32540-2_2

We are asking future generations to address the climate change and societal instability that we have caused. Each generation bears a responsibility, and my generation is waking up too late. I am not convinced Hurley et al. (2022) are correct in predicting that the younger generation will be the most resilient. This generation has, however, the possibility of learning and dealing with disasters and chaotic situations. I think we are currently working with a young generation who require support in becoming resilient. In order to cope or live, they need resilience, and the argument I make is that we need to learn how we might foster and facilitate this during education.

A year ago, a valued conference colleague who works at a prominent software development firm, remarked, "Engineering education continues to operate under the assumption that we can rectify issues. We ought to educate students on how to manage the end of the world. But I will not say this in the panel, of course." Maybe this is the problem: that we recognize the complex situation in life and in the world. However, we do not say it aloud, although there is plenty of research evidence that we will be dealing with a new type of student who is anxious about the future.

Research on emotions and climate change is fast increasing with significant findings, indicating that we are dealing with a serious problem. The quote from Hurley et al. (2022) is taken from a response to one of the largest studies on young people's anxiety for the future. This is a global survey covering young people aged 16–24 from Australia, Brazil, Finland, France, India, Nigeria, the Philippines, Portugal, the UK, and the USA, with around 1000 participants per country. The findings indicate significantly that 59% of the respondents reported that they were extremely worried, and more than 50% reported emotions like sad, anxious, angry, powerless, helpless, and guilty (Hickman et al., 2021). A Canadian study presents similar results from a survey study encompassing 1000 students aged 16–25. Similar findings indicate that more than 50% experienced feeling afraid, sad, anxious, and powerless. Nearly 80% indicate that climate change impacts their mental health (Galway & Field, 2023).

In Scandinavia, several studies show similar results that young people are experiencing the same feelings. A Norwegian study examines debate articles written by young people between the age of 13 and 21 in a period between 2016 and 2021. The study focuses on how young people express their feelings and what words they are using. Worry, fear and anxiety are the most often used combined with guilt and sadness (Andersen, 2023).

Written debates in public newspapers will represent an active and policy-oriented group of young people, as well as a public voice that ought to be heard at the political level. Social media also plays an important role of influencing attitudes and perceptions, and another Norwegian study reveals that young people are reluctant to participate in debates on climate issues and that there is a feeling of isolation or insulation (Wigum, 2024).

In a semantic study, the use of concepts used by the Danish media to describe the climate impact on mental health presents interesting results. In general, climate impact is increasingly mentioned and the psychological reaction to climate is increasing. There is a move from climate concern to examples like climate grief or climate shame. Climate grief refers to the position of the self in a state of more or less hopelessness, whereas climate shame is related to our way of living, either by feeling shame of overusing the earth's resources or not wanting to feel shame of doing it (Boas & Hougaard, 2025).

No matter what angle climate shame represents, there is no doubt that climate issues do affect people emotionally, and studies on emotions are starting to emerge (Brosch, 2021). However, research studying or measuring feelings is very difficult quantitatively but is more possible for qualitative methods. It is still in an early stage of research that calls for more knowledge and disciplines to develop methodologies (Wu et al., 2020).

I am uncertain if I am prepared to align myself with those who 'dramatically proclaim' that circumstances will deteriorate significantly. However, I do recognize that we inhabit a world characterized by an unpredictable political landscape and pressing climate challenges, juxtaposed with generations of youth that appear increasingly susceptible, with a higher prevalence of psychological diagnoses than ever before. Fear and worry appear to prevail more significantly than in my generation, where we mostly learned to keep going and persist regardless of personal desire. We discovered that motivation gradually emerged as we commenced our tasks, which gave some kind of satisfaction with our jobs. We learned to grit our teeth and do things even though we thought it was hard and often created periods of anxiety and stomachache.

2.2 Engineering, Students, and Climate Anxiety

In one of the studies on liberal art and planning students' approaches to sustainability, the findings clearly indicated that planning students (a technical, engineering discipline) had a belief in rational technological solutions, and liberal art students had emotional reactions like blame and limited engagement (Servant-Miklos & Noordzij, 2021). These are two different disciplines, and the level of knowledge of understanding climate change and having ideas of solutions are very different for the two groups. As such, one would expect that climate anxiety would not exist to the same degree among engineering students as among other disciplinary student groups; however, this is not the case. Even if engineering students do have more scientific knowledge of causes and might have solutions to problems, research still indicates that they have serious emotional worries.

In a paper by Newton and Romkey (2024), they apply the Climate Anxiety Survey (CAS) and report findings from a study encompassing 188 Canadian engineering students. Engineering students are affected by climate change although not to the same degree as other groups in the society when comparing data from the CAS (Newton & Romkey, 2024). A study surveying students from sustainability studies in three European universities in Sweden, Poland, and Finland documents that climate anxiety is present among university students (Khalaim & Budziszewska, 2024). Some perspectives from this study show that we must understand learning trajectories from an emotional point of view, and not at least that anxiety and hope can exist side by side. Another point shows that we must live with a changed world with many insecurities, among which is the existence of climate change.

An Australian study on students in a semester-long course on exploring emotional experiences of some undergraduate students in sustainability indicates that anxiety, frustration, overwhelm, guilt, grief and hope exist side by side, and that we all have to learn to live with contrasting feelings related to live-with climate change (Verlie, 2019).

> Learning to live-with climate change involves the appreciation of the entanglement and co-becoming of the human and the climatic and the affective adaptation that must accompany this. Learning to live-with climate change is therefore a process of bearing worlds, as we simultaneously become more attuned to our enmeshment with the more-than-human, mourn those relationships as they are ruptured, act-with them to cultivate the most promising futures possible, and are ourselves changed throughout the process. (Verlie, 2019, p. 752)

'Living-with' is this slow adaptation process or accommodation to the circumstances. We cannot deny climate change any longer, and we need to find ways to live-with it. If we use concepts such as 'coping' or 'living-with', it is then on the same scale. Coping represents a higher uncertainty, whereas living-with somehow signals that we are adjusting.

Living-with and being able to act also requires that engineering students to learn to address climate change in their education. A study indicates that possible factors influencing students' wishes to address climate in their careers include specific course topics, co-curricular experiences, climate knowledge, political affiliation, and, as the most significant, a personal responsibility to deal with environmental problems. This is based on a recognition of climate change as a technical issue, believing that climate change is caused by burning fossil fuels and livestock production (Shealy et al., 2021). This study clearly indicates that knowledge also plays a crucial role in emotional reactions.

A Canadian study reveals a complex relation between experiences with climate change and ethical behavior emotional response, which can be helpful in understanding and designing education (Sielmann & Giang, 2025). Another study encompassing students from Europe and South America finds that, due to their knowledge level, engineering staff and students are potentially more involved in climate activities (Ferrari et al., 2025).

There is much more literature out there that could be cited to document these points, such as the work by Andreotti that focuses on personal empowerment, the mastery of knowledge and skills, and how we can integrate them into education with a focus on the individual growth (Andreotti, 2024; Horsthemke, 2024). A recent review (based upon 40 studies) indicates that the chaos or collapse issues are highly relevant in education but are not reflected in the curriculum (Díez-Gutiérrez et al., 2024).

The literature in this field is currently exploding, and I do no justice to the field with this very brief overview. The intention here is to draw attention to an increasing issue with young students' relationship to climate change and how they react. In academia the knowledge dimension is basic and addressed in the curriculum, but what we learn from this research is that the emotional dimension calls for attention in the curriculum, which I think we need to include.

2.3 From Climate Anxiety to Chaos or Collapse?

I am a hesitant soul and need to think, document and reflect before I react. It took me some time to process the book, *Pedagogies of Collapse* by Ginie Servant (2024). Ginie has a background in philosophy and completed a PhD focused on PBL in higher education with a focus on examining the history of the topic (Servant, 2016; Servant & Schmidt, 2016). After her PhD, she came to Aalborg University as a postdoc on PBL and sustainability. We faced challenges in evaluating data from engineering students concerning their approach to sustainability.

Ginie perceived the data as indicative of an instrumental and neglectful culture, whereas we regarded it as technical strategies for sustainable development. This conflict often arises in engineering education research between sociologists and philosophers, who analyze data through a critical theoretical lens, and engineers who approach it from an innovative and pragmatic standpoint of how we can contribute. The two cultures are distinctly different, and their discourses vary significantly.

Irrespective of the discussions, Ginie's assertion is that the climate and sustainability of this planet will culminate in a collapse, necessitating the preparation of our higher education students to realize and confront this reality and to change the educational system in order to build up their resilience (Servant, 2024).

I fully understand the points outlined in *Pedagogies of collapse*; however, there is a sliding scale from chaos to collapse and with the climate situation, the political scene in 2025, and the ongoing war in Europe, we have created a world of chaos where we cannot feel safe. This calls for further anxiety. Along with climate anxiety, one of the referenced studies also includes war as a significant cause for anxiety (Khalaim & Budziszewska, 2024).

Instead of collapse, I would prefer to adopt the concept of chaos because we will be approaching turbulent scenes in the future, just like we do now. If you have experienced multiple episodes of chaos, it could lead to collapse or disorder. Chaos can be an indicator of collapse, but there is still potential for reacting and bringing chaotic incidents under control. The collapse is dramatic, with everything falling apart.

Snowden and other co-authors have developed the Cynefin framework, which provides a way for decision makers to approach various problems (Kurtz & Snowden, 2003; Snowden, 2000; Snowden & Boone, 2007), see Fig. 2.1. Decision making is one of the disciplines in leadership as decisions are part of any design or development process. As Roger Hadgraft and I have worked with the integration of complex problems into engineering education, the Cynefin framework has been very useful in helping us understand the nature of complexity compared to simple and complicated problems (Hadgraft & Kolmos, 2020).

'Simple' problems are well defined within existing disciplines and with well-known solutions. The process in the *simple* category is to sense, categorize, and

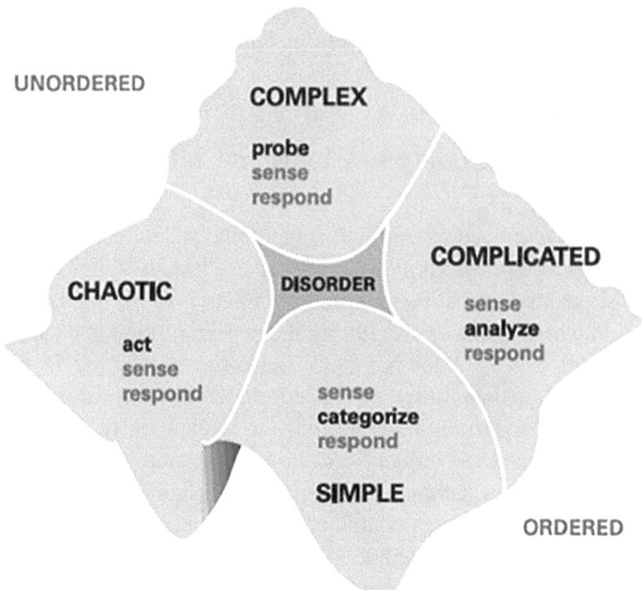

Fig. 2.1 The Cynefin framework (Snowden & Boone, 2007, p. 4)

respond. This means that you recognize the problem, categorize it within known knowledge, and respond appropriately.

The 'complicated' problem or 'domain' requires more, as there could be many solutions across existing disciplines. The process here is to sense, analyze, and respond, which means to recognize the problem and analyze it before responding. Both the simple and complicated problems belong to an ordered system to which we already know the boundaries and more or less the interrelations between disciplines and relations to practice.

In contrast to the ordered domains, we find the 'disordered' domains, which consist of complex and chaotic problems. The 'complex' problems are not well known or defined. Therefore, the process is probe, sense, and respond, which means to start out by a thorough investigation and analysis into the challenge before sensing and responding. The problem is unknown; the knowledge components to identify and analyze are unknown. This is a domain for the emergence of new knowledge, competencies, and technologies that involve disciplinary boundary crossing. Furthermore, the complexity will be on the edge of chaos or at least contain periods of chaos. However, this will be the case for all the categories—that in practice there will be overlap between simple and complicated and between complicated and complex.

There are significantly different patterns of reactions in chaotic situations, as there is no time for probing, analyzing, or categorizing. The acting is required as the first step, as we are talking about disasters like climate change, technology collapse, and terror/war situations. So, the process is to act, sense, and respond. In engineering education, there is not any one discipline that prepares students for such experiences; on the contrary, the scientific and technical curriculum is based on rationality and learning analytical and logical thinking. Maybe there will be components of risk analyses involved in some of the disciplines, hardly as an action competency but more in the subject. Not as learning of acting but more as a theoretical risk.

Chaos is frequently discussed as a phenomenon anticipated in 50 to 70 years in the future, which feels unreal to us today, but can also manifest as soon as tomorrow. The recent power outages in Spain and Portugal underscore the urgency of initiating preparations immediately. The military threat from Russia may be significantly more sophisticated by targeting infrastructure, such as severing underwater cables, as evidenced by the attack on the pipelines transferring Russian gas to Germany, or the ever-increasing cyberattacks, such as drones disturbing flight traffic or jamming the traffic signals. I believe that right now we are in a situation where only fantasy creates boundaries for what can happen.

We cannot move forward if we end up feeling anxious and afraid because the world does not feel safe. Learning is disturbed by a high degree of anxiety, and how can the educational system respond? We are dealing with some dilemmas here as we cannot learn when we feel anxious, but we also need resilience to overcome anxiousness. Chaos has always existed and the Western world has not seen war since the Second World War followed by the conflicts in Yugoslavia and Ukraine. So, for my generation and for the younger generations, this is new.

In the smaller numbers, we might find some people who actually thrive in chaos. That there will be different disruptions and periods of discontinuity to allow for new ideas and new systems. However, what we see today is more than minor periods of chaos, and the concept of living-with climate change or chaos might be a strategy that encompasses both concerns and hope.

We need to meet challenges, to learn to take risks, and to overcome obstacles by identifying individual and collective solutions. In education we need to challenge the students both at the knowledge and social levels, and the learning experience is a powerful methodology. The students can experience how to deal with situations where there are no constraints and no visible boundaries: How will they act? What kind of constraints are important in the first place to try to frame the problem?

Risk-taking is needed in this process. If you don't know what's going on, it isn't possible to act on safe knowledge ground. You will need to take risks to encapsulate or frame the problem. Risk-taking is challenging for both individuals and teams. Risk-taking can be based on knowledge; however, the courage to carry out the acting is emotional, as indicated in the previous sections. Students need to learn to act and think of ethical consequences for society.

2.4 Response from Engineering Education

For some years we have worked with the integration of complex problems in engineering education and the ways we could develop the curriculum (Hadgraft & Kolmos, 2020). We were inspired by Snowden and Boone's framework, and with this as a starting point, we analyzed and framed the engineering curriculum by differentiating between dimensions such as those illustrated in Fig. 2.2:

- Disciplinary silos versus cross-disciplinary approaches.
- Simple and complicated problems versus complex and chaotic issues.
- Technical knowledge versus context, systems, sustainability, and values.
- Individual modules versus curriculum models.

The established pedagogical strategies were categorized between teacher-centered and student-centered learning approaches, encompassing instructional textbook learning, active learning in many media, and culminating in professional competencies and individual learning trajectories. In the article, Roger Hadgraft and I highlight the last level encompassing professional abilities and personal growth trajectories—an emerging trend that remains to be completely realized and which I will continue to develop in the chapters on generic competencies (Chap. 6) and boundary crossing (Chap. 8).

We identify active learning and especially problem- and project-based learning as core pedagogies, which can be a response to complex and chaotic situations. This is in line with a recent book on education and learning for sustainable futures, a thorough review of the understanding of sustainability and not at least the educational approach where the future is expected in direction of active and social engagement (Macintyre et al., 2025).

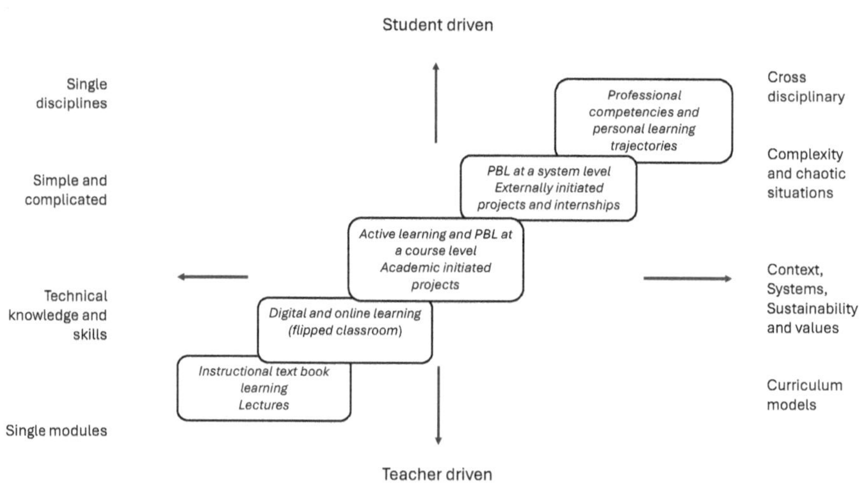

Fig. 2.2 Emerging learning environments (Hadgraft & Kolmos, 2020)

In the academic literature, there are plenty of articles and books calling for change in engineering education. In an editorial for a special issue on sustainability for the Journal of Engineering Education, Martin et al. call for change in engineering education due to climate change. They call for four areas (Martin et al., 2022, p. 740):

(1) understand how climate and sustainability are linked to engineering design
(2) incorporate a wide range of disciplines into engineering solutions
(3) understand the ethics and justice dimensions of engineering
(4) listen to and collaborate with diverse communities.

Essentially, they call for sustainability in engineering design, interdisciplinarity, ethics, and transdisciplinarity. Caroline Baillie and Paul I. Kadetz from the University of San Diego have contributed with a book entitled 'Reimagining Engineering Education' (Baillie & Kadetz, 2024). Caroline Baillie's primary emphasis is to integrate social justice in engineering education, and she has successfully initiated a social justice movement while developing new courses and programs within a framework of sustainability and social justice. A significant theme of her book can be succinctly encapsulated by this quotation:

> Thus, reimagining engineering education will not merely involve imparting additional or alternative concepts. So, if we are to reimagine engineering education, it is not going to be a simple case of teaching more, or different, concepts. We need to ultimately rethink our thinking and question the ways of knowing and being at the very core of engineering. We need to reevaluate our values, to consider what is being engineered and for whom. However, when serious transformations occur, there is strong resistance to change. This is especially true when those in power benefit the most from the current system: be it company engineers and their CEOs, or educators and their administrators receiving funding from those same engineering companies and individual donors. ((Baillie & Kadetz, 2024, p. 3)

The essence of this quote is that engineering education does not change by incorporating additional competencies or courses into the curriculum; we need a fundamental transformation in our conceptualization of engineering and its societal role, not more of the same. Nevertheless, this shift in thought and reevaluation of values will encounter opposition. The sociocultural drivers and mechanisms are entrenched inside a systemic and intricately interconnected power structure, which cannot be dismantled just through curricular approaches. Despite the shift of engineering education from a predominantly lecture-based curriculum to a more active learning paradigm, it continues to adhere to the same societal market-oriented principles, prompting one to doubt the truthfulness of this transformation.

Baillie and Kadetz observe that the transformation in engineering education emphasizes learning styles, teamwork, and design, prioritizing micro-ethics above a critical macro-ethics that fosters social awareness. Micro-ethics pertains to individual responsibility, while macro-ethics is fundamentally rooted in critical societal awareness and social movements advocating for social justice (Baillie & Kadetz, 2024). They touch on a core problem for engineering as engineers stand back from politics. Engineering is a rational and logical dominated domain that drives its arguments on facts and evidence. This is in contrast to the political scene, which is driven by power, negotiation, and compromises leading to social regulation. Two very different sets of values.

In a recent publication that I co-authored, we attempt to forecast the future by rethinking engineering education to tackle human concerns (Habbal et al., 2024). The conclusion emphasizes nine domains for transformation:

1. Engineering curricula should be design-oriented and interdisciplinary, with a focus on solving open-ended, complex, human challenges.
2. Engineers must adopt socio-cultural environmental and innovative mindsets.
3. Interdisciplinary knowledge is the cornerstone for solving these complex human challenges—excellence in a single discipline must not be the only focus.
4. Learning environments must facilitate learning as a social process.
5. Experiences, variation, and reflection should be practiced throughout the curricula.
6. Students and teachers need generic and meta-competencies to work across interdisciplinary boundaries.
7. Students must be encouraged to create their own lifelong learning trajectories.
8. Disciplines must embrace interdisciplinarity.
9. Digitalization is changing the engineering profession and our learning environments.

Baillie and Kadetz integrate social justice into engineering education, which encompasses not only a socio-cultural and environmental perspective but also issues of poverty and equitable distribution between the Global North and South. Ginie Servant underscores the importance of self-awareness and the cultivation of coping methods in a world plagued with significant threats that alarm, if not terrify, the youth regarding their future (Servant, 2024). In our book, this connection is lacking as an explicit argument and requirement (Habbal et al., 2024). I am convinced that all authors of our book agree, yet it is not expressly articulated.

We also didn't address the integration and consequences of chaos. Conversely, I was among the authors asserting that we cannot resolve chaos in education. Perhaps we were wrong, and maybe we were also about to integrate chaos in engineering education, but we did not do it. It has taken some time to comprehend that we may have to alter our ideas and go one step further from complexity to also include chaos. Even if complexity and chaos do have overlapping features, learning-wise we have very different learning phases. Complexity does require investigation across the disciplinary boundaries. Chaos does require action.

2.5 University Modes

I believe what we are calling for goes beyond the critical university modes that have been presented throughout the last 20–30 years. In an article co-authored with Andrew Jamison and Jette Holgaard, we identified three fundamental university modes: the academic mode, emphasizing theory; the market-driven mode,

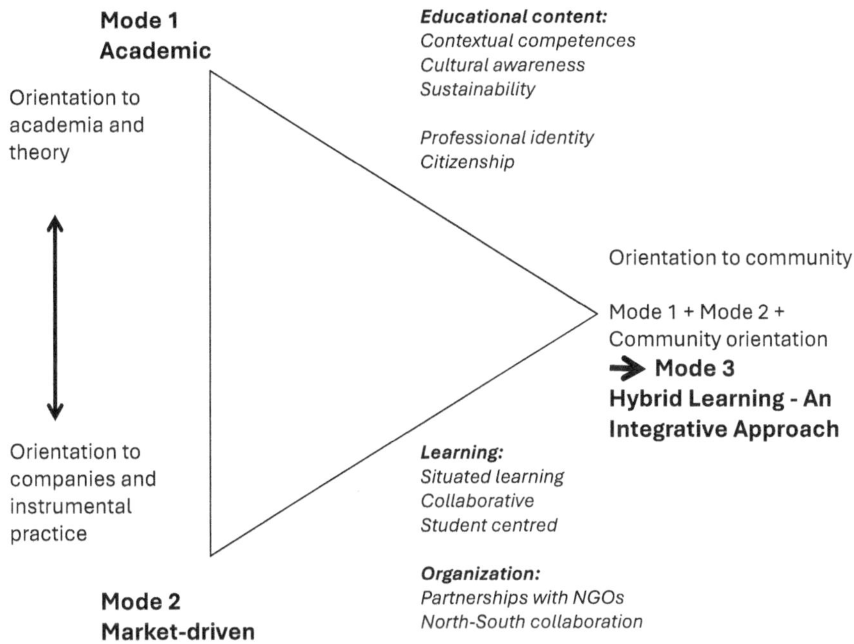

Fig. 2.3 Three university modes (Jamison et al., 2014)

concentrating on competencies; and the critical societal mode, as illustrated in Fig. 2.3 (Jamison et al., 2014).

These modes exemplify an idealized and more theoretical perspective. Most institutions incorporate aspects of all three types, albeit they predominantly emphasize either the academic or market-driven modes. In an academic framework there will be segments of market-oriented courses, and conversely in the market-oriented framework there will be an emphasis on theoretical academic knowledge and its practical application from a market-driven viewpoint. The academic mode is constrained by diminished academic independence due to restricted research resources and an emphasis on externally sponsored research, which narrows research objectives to align with project goals.

The mode 3 approach remains underdeveloped in engineering education. It occurs as part of specific courses but not on a full scale. Even mission-driven universities, a recent trend in Europe, use market-driven ideas. We must still formulate a mode 3 mode, and the past decade has undoubtedly necessitated a reevaluation of mode 3 values. However, in mode 3, even if we talk about citizenship, cultural awareness, sustainability as the content, and student-centered, collaborative situated learning, the interdisciplinary, transdisciplinary, and the chaos dimensions are missing.

These three modes are equivalent to the three understandings of competencies, which Ron Barnett developed in the late 1980s and early 1990s, see Table 2.1 (Barnett, 1990). The three categories are: operational competence, academic

Table 2.1 Modes of competences (Barnett, 1994, p. 179)

	Operational competence	Academic competence	Lifeworld becoming
1. Epistemology	Know-how	Know that	Reflective knowing
2. Situations	Defined pragmatically	Defined by intellectual field	Open definition (with use of multiple approaches)
3. Focus	Outcomes	Propositions	Dialogue and argument
4. Transferability	Meta operations	Metacognition	Metacritique
5. Learning	Experiential	Propositional	Meta-learning
6. Communication	Strategic	Disciplinary	Dialogical
7. Evaluation	Economic	Truthfulness	Consensus
8. Value orientation	Economic survival	Disciplinary strength	The 'common good'
9. Boundary conditions	Organizational norms	Norms of intellectual field	Practicalities of discourse
10. Critique	For better practical effectiveness	For better cognitive understanding	For better practical understanding

competence, and lifeworld becoming. Operational competence is what Jamison et al. (2014) call the market driven mode with a focus on know-how, pragmatism, outcomes, meta operations, experiential learning, strategic communication, economic evaluation, organizational norms—with the purpose of effectiveness. The academic competence is based on epistemology and is related to situations defined by the intellectual field, propositions, metacognition, propositional learning, disciplinarity, truth, disciplinary strength, intellectual norms—with the purpose of a better cognitive understanding.

The lifeworld becoming is reflective knowing, open definitions, dialogue, metacritique, meta-learning, dialog-based, consensus, the common good, the practicalities of boundaries/discourses—all with the purpose of a better practical understanding. This was developed 20 to 30 years before we discussed climate change and complex problems in education. Therefore, it is also prior to the agenda of dealing with chaos. However, Barnett points in directions of new types of learning and values that are based on critical discourse. These might not belong to the competency discourse but to a much broader understanding of the role of higher education.

Barnett has the vision for a third dimension. During the 1990s, pressure from governments increased to promote higher education in the direction of more market-oriented competencies. Employability came into the vocabulary of higher education and was often implied as moving away from the more general education, also understood as Bildung. Bildung is the overall formation to act as a citizen working for the progress of society with respect to history and the individual. In the employability literature, Yorke has made the concept broader, seen as

> a set of achievements—skills, understanding, and personal attributes—that makes graduates more likely to gain employment and be successful in their chosen occupation, which benefits themselves, the workforce, the community, and the economy. (Yorke, 2004, p. 410)

Even if there are no black or white boxes in reality and the definition by Yorke is broader and covers the community, it is still within a narrower market-driven approach that does not question the societal values, including meta-learning and dialogue.

Another important reason to integrate Barnett is that in the book *Limits of competence*, he criticizes the problem-based approach for being too instrumental and short-sighted (Barnett, 1994). On the other hand, Ginie Servant in her book *Pedagogies for collapse* points at potential in the problem-oriented project work for educating young people to resist and cope with a possible collapse. It reminds us that there is variation in the learning outcomes in PBL studies ranging from disciplinary learning outcomes to broader outcomes of societal understanding and self-development. I will return to this in Chaps. 3, 4, 5, and 6. I will argue that the problem-oriented project work and other active learning methodologies can be a base for responding to the contemporary anxiety and well-being issues of today. This is no longer enough, however. The curriculum has to be combined with elements of learning to act and learning to reflect knowledge and identity as whole persons embracing both knowledge, emotions, values and social collaboration.

Upon critical reflection of my previous work on PBL, I aspire to further this domain by highlighting a new phase of PBL. My main message is that we need variation and inclusion of curriculum elements for individual and collective resilience to be strong enough to effectively navigate around in the chaotic world. The PBL that we developed 50 years ago is not outdated; however, in many cases it has become too uniform and instrumental as the practices in engineering education have become too much alike. It only transforms into another conventional pedagogical approach. It is essential to integrate personal and societal values into the learning of technology and incorporate personal and societal values into engineering education. A condition for this is that we can foster and encourage academic staff to embrace a broader learning philosophy.

I will basically analyze the development of PBL, understood as both problem-based learning and project-based learning, to envision new developments of this pedagogy. To change to versions of PBL might be a huge step for many institutions; however, it is not the PBL that is the end goal, but indeed to help equip young people to live-with and live-in a changing world, which at times will be chaotic.

2.6 Reflection 2: Living-with Chaos

During the last five years, I have personally experienced a growing concern and worry for the world. Not only climate issues but indeed also the endless and meaningless wars at the global level. This lack of global communication and dialogue and lack of intention for understanding each other might give a feeling of depression and powerlessness at a personal level. Starting to review the literature on young peoples' anxiety for the future was frightening as it involves the future of mankind, and I feel responsible to raise the flag and rethink my generation's contribution to the change of engineering education. There is much more literature out there on this topic to which I could have referenced, but I feel I have made the point.

My generation has also tried to indicate alternative directions in engineering education. We have developed models, conducted studies of existing educational practices, and provided data. We have looked at new learning methodologies and alternative university modes to further educate both work and life. We have been involved in educational changes and have created lighthouses of pedagogical practices, of which some will be described in the coming chapters. The question we have is: Was this enough, worth the effort, or just a base for further development?

This is what I aim to address in the next section of this book. What kind of changes did we make to the curriculum and what can we learn from our experiences with past transformations to problem- and project-based learning systems? During the 1960s and 1970s there were a lot of good intentions, and maybe we overlooked some aspects of learning. I belong to a generation that tried to build up a critical mode by the development of PBL and the intentions were good, but maybe we forgot them in our daily practices. Or did the good intentions vanish during phases of organizational rationality and effectivization? And if the latter, how can we bring the potential back?

Some institutions are at the edge of trying to apply more active learning methodologies. And for those institutions, this book might provide a shortcut to a modernized curriculum that can emancipate and empower students to live-with chaos and complexity in the world of today.

References

Andersen, I. V. (2023). «Jeg er redd for fremtiden»: Unge, bekymrede stemmer i klimadebatten. *Bergen Language and Linguistics Studies*.

Andreotti, V. D. O. (2024). The task of education as we confront the potential for social and ecological collapse. In *Education, the environment and sustainability* (pp. 7–22). Routledge.

Baillie, C., & Kadetz, P. I. (2024). *Reimagining engineering education*. Springer.

Barnett, R. (1990). *The idea of higher education*. Open University Press.

Barnett, R. (1994). *The limits of competence: Knowledge, higher education and society*. Open University Press.

Boas, K., & Hougaard, T. T. (2025). Klimasorg eller klimaskam? Psykologiserende klimakomposita som led i en social positionering. *NyS, Nydanske Sprogstudier, 67*, 7–35.

Brosch, T. (2021). Affect and emotions as drivers of climate change perception and action: A review. *Current Opinion in Behavioral Sciences, 42*, 15–21.

Díez-Gutiérrez, E.-J., Trujillo Vargas, J. J., Palomo-Cermeño, E., Perlado-Lamo de Espinosa, I., García-Salas, L.-M., Romero Acosta, K., Mateos-Toro, L.-M., & Pérez-Robles, A. (2024). Educating in and for degrowth: Training future generations to prevent environmental collapse. *Sustainability, 16*(21), 9210.

Ferrari, E., Whitmarsh, L., MacCann-Alfaro, N., Cáceres-González, C., Melara, E., Hernadez-Gallego, E., & Ruiz, C. (2025). Can climate anxiety and awareness drive academic behaviour in higher education? In *Education for sustainable development: The contribution of universities* (pp. 59–74). Springer.

Galway, L. P., & Field, E. (2023). Climate emotions and anxiety among young people in Canada: A national survey and call to action. *The Journal of Climate Change and Health, 9*, 100204.

Habbal, F., Kolmos, A., Hadgraft, R. G., Holgaard, J. E., & Reda, K. (2024). *Reshaping engineering education: Addressing complex human challenges*. Springer Nature.

Hadgraft, R. G., & Kolmos, A. (2020). Emerging learning environments in engineering education. *Australasian Journal of Engineering Education, 25*(1), 3–16. https://doi.org/10.1080/22054952.2020.1713522

Hickman, C., Marks, E., Pihkala, P., Clayton, S., Lewandowski, R. E., Mayall, E. E., Wray, B., Mellor, C., & Van Susteren, L. (2021). Climate anxiety in children and young people and their beliefs about government responses to climate change: A global survey. *The Lancet Planetary Health, 5*(12), e863–e873.

Horsthemke, K. (2024). *Education, the environment and sustainability*. Routledge.

Hurley, E. A., Dalglish, S. L., & Sacks, E. (2022). Supporting young people with climate anxiety: Mitigation, adaptation, and resilience. *The Lancet Planetary Health, 6*(3), e190.

Jamison, A., Kolmos, A., & Holgaard, J. E. (2014). Hybrid learning: An integrative approach to engineering education. *Journal of Engineering Education, 103*(2), 253–273. https://doi.org/10.1002/jee.20041

Khalaim, O., & Budziszewska, M. (2024). "It should not only be technical education." Students' climate anxiety experiences and expectations toward university education in three European universities. *The Journal of Environmental Education, 55*(4), 308–323.

Kurtz, C. F., & Snowden, D. J. (2003). The new dynamics of strategy: Sense-making in a complex and complicated world. *IBM Systems Journal, 42*(3), 462–483.

Macintyre, T., Tilbury, D., & Wals, A. (2025). *Education and learning for sustainable futures: 50 years of learning for environment and change*. Taylor & Francis.

Martin, M. J., Diem, S. J., Karwat, D. M., Krieger, E. M., Rittschof, C. C., Bayon, B., Aghazadeh, M., Asensio, O., Zeilkova, T. J., & Garcia-Cazarin, M. (2022). The climate is changing. Engineering education needs to change as well. *Journal of Engineering Education, 111*(4).

Newton, H., & Romkey, L. (2024). Examining climate anxiety and sustainability engagement in the undergraduate engineering student population. In *2024 ASEE Annual Conference & Exposition, Portland, Oregon*.

Servant, V. (2016). *Revolutions & re-iterations: An intellectual history of problem-based learning*. Erasmus University College (EUC).

Servant, G. (2024). *Pedagogies of Collapse: A hopeful education for the end of the world as we know it*. Bloomsbury.

Servant, V. F., & Schmidt, H. G. (2016). Revisiting 'Foundations of problem-based learning: Some explanatory notes'. *Medical Education, 50*(7), 698–701.

Servant-Miklos, G., & Noordzij, G. (2021). Investigating the impact of problem-oriented sustainability education on students' identity: A comparative study of planning and liberal arts students. *Journal of Cleaner Production, 280*(2), 124846.

Shealy, T., Katz, A., & Godwin, A. (2021). Predicting engineering students' desire to address climate change in their careers: An exploratory study using responses from a US National survey. *Environmental Education Research, 27*(7), 1054–1079.

Sielmann, C., & Giang, A. (2025). Pilot study: Student experience of climate change anxiety and student inclination toward ethical behaviour. In *Proceedings of the Canadian Engineering Education Association (CEEA)*.

Snowden, D. (2000). Cynefin: A sense of time and space, the social ecology of knowledge management. In C. Despres & D. Chauvel (Eds.), *Knowledge horizons: The present and the promise of knowledge management*. Butterworth Heinemann.

Snowden, D. J., & Boone, M. E. (2007). A leader's framework for decision making. *Harvard Business Review, 85*(11), 68.

Verlie, B. (2019). Bearing worlds: Learning to live-with climate change. *Environmental Education Research, 25*(5), 751–766.

Wigum, K. (2024). *Klikk for klima: Sosiale mediers rolle i formingen av unge voksnes klimaoppfatninger og-handlinger*. NTNU.

Wu, J., Snell, G., & Samji, H. (2020). Climate anxiety in young people: A call to action. *The Lancet Planetary Health, 4*(10), e435–e436.

Yorke, M. (2004). Employability in the undergraduate curriculum: Some student perspectives. *European Journal of Education, 39*(4), 409–427.

Chapter 3
Variations in Institutional PBL Models

3.1 Introduction

In the previous chapters, I have described some aspects of the chaotic world we are living in. I am not the only academic who are worried. Along with thousands of other academics, Baillie, Kardetz, and Servant have posed the following questions (Baillie & Kadetz, 2024; Servant, 2024):

1. How can higher education react to questions of climate change, sustainability, and green innovation?
2. How can we prepare our youth to live in an unpredictable world as described in Chap. 2?

Academics are eager to find solutions and solve contemporary questions, and answers have pointed toward the directions of integrated curriculum, problem-based learning, and project-based learning (PBL) in both pure and combined forms. Similarly, Conceive, Design, Implement, and Operate (CDIO), active learning forms and, most recently, challenge-based learning (CBL) in engineering education have all been among the educational approaches tested during the last 30 years (Crawley et al., 2014; Dikilitaş et al., 2025; Drake & Burns, 2004; Felder et al., 2015; Kolmos & de Graaff, 2014). In Chap. 5, I will expand on these current educational trends.

Furthermore, resilience is a core quality in coping with disasters or chaos. As a concept, resilience has many meanings, although the most common is that it involves coping with unexpected and potential unpleasant situations, and learning to adapt to change and uncertainties (Hunsu et al., 2021; McEwen, 2011). To become more resilient, one must face different risky and unexpected situations, and experience cognitive and emotional reaction patterns to develop strategies. This can happen in more student-centered educational models where students are the drivers of their

A. Kolmos, *Engineering Education for Interdisciplinary Futures*, Emergent Directions in Engineering Education 2,
https://doi.org/10.1007/978-3-032-32540-2_3

own learning and experience different problems as well as ways of learning and reflecting.

During the last 40 years, there has been a broad move to more active learning both in engineering education and in higher education in general. In engineering education, active learning was promoted with the existing teaching and learning regime of traditional classroom teaching and lecturing with some problem-solving tasks (De Campos et al., 2012; Prince & Felder, 2006; Rugarcia et al., 2000). The same counts for PBL, which has had a significant impact on engineering education (Chen et al., 2020). PBL, in particular, has gained ground in engineering education, and in the next chapters, I will emphasize the variations and possibilities that this pedagogical approach offers and the disadvantages of pedagogical stagnation that can occur in moving to new teaching and learning systems.

I have always argued for a systems approach as the optimal solution, understood as a systems approach with variation in the teaching and learning methodologies. Curriculum change can be carried out with both top-down and bottom-up strategies (Kolmos, 2017). Top-down is the management and political level initiating and making decisions. Bottom-up is the academic staff who experiment with new teaching and learning methods, which will be the inspiration for others to try out new methods. Ideally, it should be a combination, as top-down strategies can meet resistance among academic staff and bottom-up can lead to a messy curriculum without an overview for the students. System approaches require involvement at all levels at an institution.

However, if systems are stiffening up and do not allow for variation in the learning and teaching methodologies, then we have just created new regimes that might not be optimal in contributing to creating resilience and learning to live-with chaos. Maybe PBL lost its potential for this, or maybe it's a platform for further development.

3.2 What Does PBL Mean?

The confusion has always been there in the literature and was reinforced when Aalborg University (AAU) during the 1990s declared that their pedagogical model was 'problem-based learning combined with project working'. The acronym of PBL is messy. The meaning of 'B' and 'L' are 'Based Learning'; it's the 'P' that creates confusion:

- Problem
- Project
- People
- Play

- Process
- Product

Maybe all these suggestions are the 'P' as they overlap. The US K-12 education applies PBL as acronym only for project-based learning. "*Project Based Learning (PBL) is a teaching method in which students learn by actively engaging in real-world and personally meaningful projects*" (PBLworks, 2025). In literature, there are other acronyms trying to create a distinction between the case-based PBL and the project-based PBL, which are the two original institutional models, e.g. Maastricht and Aalborg Universities. There is also a PjBL acronym, which stands for project-based learning (Graham, 2010).

The idea of underlining the project and the problem separately might make good sense. The problem is the trigger for learning, and especially when students are working with authentic or real-life problems their interests and motivation increase. The project is the organization of the learning process, and it does create the framework for the competencies. These are among the core reasons for changing to a PBL curriculum. At Aalborg University there were attempts to acronymize it as 'Project organized and problem based learning' (POPBL) (Lehmann et al., 2008; Moesby, 2008). The acronym POPBL never became a widespread international trend nor internally at AAU. During the 1980s, it was known as project work and problem-oriented project work.

In this book, PBL denotes problem- and project-based learning as combined curriculum models and learning principles. Actually, the problem is emphasized twice as all projects will start out with a problem to solve. This is embedded in the definition of a project, which is a unique endeavor for a specific purpose (Algreen-Ussing & Fruensgaard, 1992). However, in many project practices, the problem is given to the students like a task, and therefore it makes sense to emphasize it twice.

The curriculum and classroom practices are so rich and so diverse that it would never be possible to name each single practice, and this is not the intention with PBL pedagogy. In my view, the intention with PBL is to motivate students' deeper and critical learning of contemporary societal and scientific challenges and empower students' learning of knowledge, competencies, and resilience to participate, act in, and construct sustainable futures.

PBL is a holistic pedagogy addressing students' learning, their interaction with society, and the collaborative learning environment that can be combined with individual growth. However, if PBL does not have variation, it becomes routine just like the lecturing system. In many textbooks the problem is regarded as a task, and there is a need to emphasize the openness in the problem approach. As in all other aspects of life, the concept of what a problem should be perceived as will vary depending on the situation, and there is a continuum ranging from more narrow disciplinary difficulties to open or ill structured society problems.

The PBL concept can be manifested at three levels: practice, models, and theory. So, when people ask the standard question, "What is PBL?" my response has often been that it depends on what level they want to have the answer. I have been to conferences where the topic of discussion was whether it would be called PBL if the teams had four students instead of twelve. If the tutor was a discipline expert, could it be referred to as PBL? Both queries are nonsensical in my opinion because PBL is largely about concepts, models, and learning activities.

We cannot develop new pedagogies or educational models based on features like team size or whether the problems are defined by the teacher or the students. Practice requires variation, and variation or spectra are easier to conceptualize within frameworks, models, or theories. PBL is not a recipe, it is a learning strategy based on the learners' intrinsic motivation and extrinsic motivation by social and societal challenges as the drivers. It can contribute to a curriculum where students learn to deal with the societal complex and chaotic problems and form the base for individual and collective growth.

3.3 The Origins and the Radical Theories

The problem-oriented project work was institutionalized during the 1970s in Denmark. This was a period of possibilities. Societies had recovered after the two World Wars, and the spirit of freedom and possibility was in the air. The student movements were strong all over the Western world. Optimistic criticism of capitalism was part of the daily rhetoric and visions for another society with peace, renewable energy and alternative ways of living were fundamental values for most of the students at that time. We learned to raise our voices, and to be critical was a must in that culture. The role of the universities changed, becoming part of the development of participatory democracy (Wright, 2025).

Several universities were established during the 1960 and 1970s (Pellew & Taylor, 2020). McMaster, Maastricht, Roskilde, and Aalborg Universities are all famous and were 'born' with new pedagogical ideas and teaching and learning approaches. The origins have been studied in a PhD thesis by Ginie Servant where the deeper historical details are explained (Servant, 2016; Servant-Miklos et al., 2019). Servant cites the reform universities as the foundation for these pedagogies.

The history of Denmark's reform universities has been told many times in the literature. Ginie Servant's PhD thesis, *Revolutions & re-iterations: An intellectual history of problem-based learning* is a scholarly introduction and examines the history of McMaster, Maastricht, Roskilde, and Aalborg Universities. It is worth reading (Servant, 2016).

The theoretical ideas of the new pedagogy in Denmark emerged from the German conceptions, such as experiential learning that Negt and Kluge developed as an emancipatory pedagogy for the working class (Kluge & Negt, 2016). In the early days of Aalborg University, only Negt and Kluge, and critical theory by the Frankfurter School were referred to as the origin for the problem-oriented project work (Wiggershaus, 1994). Scholars from the Chicago group, such as Dewey and Kilpatrick, who had developed concepts of democracy and project work years earlier, were also referenced (Beyer, 1997; Dewey, 1958; Kilpatrick, 1918). Additionally, Brazilian Paulo Freire is often mentioned as source for inspiration (Freire, 1973). These writers were highly recognized as the proponents of a novel educational approach during the reform university era for democracy and more equal rights in society.

In the Danish scene, it was Knud Illeris, Jens Berthelsen, Sten Cloud Poulsen, Eva Hultengreen, Karen Borgnakke, and Jes Adolpsen who were among the leading researchers formulating these new pedagogical ideas. Berthelsen, Illeris, and Poulsen write in a book from 1977: "*There is nothing socially changing or revolutionary in the working method itself. What is crucial lies in the specific working process and the specific project content in which the working method is realized in each case* [Translated]" (Berthelsen et al., 1977, p. 17). This is a rather important point as they did not see the critical part in the collaborative project work itself, but in the problems that the students were working on.

The core principles—beside the teamwork in the projects—were (Berthelsen et al., 1977; Illeris, 1974):

1. Problem orientation involving an external collaboration with society and the possibility to relate own experiences to a more overall context, which relates to the principles of exemplarity.
2. Participant directed learning to emphasize their role as an active investigator of own and societal conditions, and that the participants had to see the problems as experienced problems. It is also called the psychological selection of content (Hultengren, 1981).
3. Teams or groups that form the basis for the projects.
4. The relation between theory and practice where the role of the academic teacher is to frame the problems in a societal understanding.
5. Interdisciplinarity—as the problem would determine the disciplines involved.
6. Exemplarity—which might have different meanings but in the original version, the problem should be exemplary for social imagination or that the problem was exemplary for the overall learning outcomes.

The researchers at that time comprehended the impact on society as both the change for the stakeholders involved and the development of the students' own societal awareness. This was in line with the theories of Negt and Kluge who formulated the need for emancipatory development of the political consciousness among working class people by relating experiences and political theory (Kluge & Negt, 2016; Negt, 1974). It was the hope that this could create political revolution which, seen as an after-reflection, is a naive political imagination and never happened; however, the pedagogy turned out to be efficient in integrating relevant competencies into the curriculum such as teamwork and project management that were required by society.

3.4 Comparing Institutional PBL Practices

The reform universities share many common experiences and even if there is an individual history related to each of them; it makes sense to compare some of the institutional PBL curriculum practices.

Reform universities are called as such because they were all granted or established with a new student-centered curriculum approach at a system level. Essentially, at the time, there were two distinct practices emerging in a variety of disciplines. As previously stated, societal developments necessitated new knowledge and skills, and several of the new universities established during the 1960s began with new pedagogical approaches and educational models. McMaster University was the first to coin the term problem-based learning and to incorporate PBL principles into a new medical curriculum (Neufeld et al., 1981; Neville et al., 2019). Maastricht University in the Netherlands, Linköping University in Sweden, and Newcastle University in Australia have all established PBL curricula in their medical schools, following McMaster's lead.

McMaster University, followed by Maastricht, organized a curriculum with thematic blocks of approximately six weeks, in which the students analyzed and solved problems described in cases. Each block period covers a specific theme, such as a group of complaints or organ systems and consists of a series of cases. A case is initiated by a problem in order to trigger the student's learning with the understanding that the problem does not need to be solved; it just serves to initiate the student's curiosity (Norman, 1998). The students can organize the analyses of the cases using a procedure called the Seven Jumps method (Schmidt & Moust, 2000b). The first objective is to define learning outcomes that guide students' learning. The typical study group (8–12 students) meets once or twice a week. In the study group each student presents and discusses their work, which is discussed and the group debates who should continue with what tasks. Research has shown the effectiveness of the self-directed learning process (Yew & Schmidt, 2009). The role of the tutor is to attend the meetings and to facilitate the group's collaboration and internal communication.

The Danish problem-oriented and project-based model was used at Aalborg University and Roskilde University in Denmark, as well as Bremen University in Germany during its early years. The establishment of these universities was based on three types of political arguments: (1) a need for new knowledge and skills from the labor market; (2) the study programs were too fragmented and disconnected from the outside world; and (3) there was a need for more democracy and student influence in the studies.

The Aalborg and Roskilde models are problem-based by definition (or problem-oriented as it was originally called) and project-organized (Illeris, 1974). For each semester, there is a theme in which the students either identify problems, or the facilitators help to scope the problems. In principle, every project starts with a problem. The students form teams of 5–8 students to analyze and solve the problem, and at the end they write and submit a project report.

Normally, the projects run for a semester (half year) allowing students to work on more comprehensive problems, and if they work on authentic problems then this can increase their motivation. Working on analyses and solutions to problems frames the learning of competencies to understand societal needs for new innovations. The collaborative part frames the learning of collaboration and project management, e.g. how to start, carry out, and finalize a project. These competencies have been required by industry for a long time (Bitsch Olsen & Pedersen, 2005; Kjærsdam & Enemark, 1994; Kolmos et al., 2004a, 2004b).

By comparing the learning principles by Barrows and Illeris, there are several overlaps as seen in Table 3.1. The problem is the start of the learning process, and there is a student-centered, participant-directed or self-directed learning approach and student groups or teams are common. As a learning principle, it is the same, but who formulates the problems is different as the teachers design problems for the cases in the McMaster and Maastricht models, whereas there is more room for students to formulate the problem in the Aalborg and Roskilde models.

The Danish models also highlight differences in their emphasis on interdisciplinarity and exemplary learning. Interdisciplinarity is, in principle, an integral aspect of Danish traditions. The interdisciplinary aspect proved challenging to implement, and in engineering education it was often limited to fostering a broader disciplinary or contextual perspective during the initial phases of problem analysis for all students. I will go deeper into interdisciplinarity in Chap. 7.

Exemplary learning is a German concept that highlights experiential learning and the significance of grasping fundamental principles over mere rote memorization of facts. Students must understand the principles and underlying concepts, allowing them to apply their knowledge across diverse contexts and engage in critical thinking. Wagenschein promoted active engagement, critical analysis and the cultivation of problem-solving abilities (Korsgaard, 2019; Wagenschein & Horton-Krüger, 2012). Wagenschein also advocated for the use of exemplary problems to direct learners in their discovery and comprehension of fundamental principles.

Klafki advanced a novel perspective on Bildung and termed the formal Bildung perspective. This approach necessitates that the curriculum remains open and that students engage with problems relevant to their experiences (Klafki, 2007). Klafki emphasized critical-constructive teaching and learning, highlighting the necessity for educational content to be selected and organized to facilitate learners' comprehension of essential concepts, the development of critical thinking skills, and an understanding of societal issues.

Table 3.1 Original learning principles at the PBL universities (Anette Kolmos & Erik de Graaff, 2014, p. 145)

McMaster and Maastricht Universities Problem-based learning	Aalborg and Roskilde Universities Problem-oriented and project-organized learning
• Problems form the focus and stimulus for learning • Problems are the vehicle for development of problem-solving skills • New information is acquired through self-directed learning • Student-centered • Small student groups • Teachers are facilitators/guides (Barrows, 1996)	• Problem orientation • Interdisciplinary • Exemplary learning • Participant-directed • Teams or group work (Illeris, 1974)

Negt and Kluge advanced the concept of emancipatory pedagogy, positing that awareness of societal issues would influence political awareness (Illeris, 1974; Kluge & Negt, 2016; Negt, 1971). Evidence indicates that engineering students and graduates from a systemic PBL curriculum demonstrate a greater awareness of societal and sustainability issues compared to their counterparts from non-systemic PBL curricula; however, this is not necessarily what was meant with political awareness during the 1970s (Kolmos et al., 2021).

Exemplarity operates in connection to practice that is relevant to core curricular areas and the student. In many ways, the German tradition has been a frontier for PBL on a larger scale. The student-centered, self-directed, and active learning approaches match with the PBL principles, as does the incorporation of societal and practical concerns into the curriculum. Furthermore, critical thinking and encouragement to explore and find problems, relevant information, and answers to problems are all consistent with the PBL principles.

The Brazilian influence began with Paulo Freire, who advocated for a social approach to learning (Freire, 1973). Freire had a much bigger political agenda and approached education as the possible emancipation and empowerment for the oppressed classes in society. He also argued for dialogue and problem-orientation with a new role for teachers to guide and learn together with students.

The learning theories align also with the Chicago group, such as Dewey and Kilpatrick, who developed ideas about project work and democracy (Dewey, 1903; Kilpatrick, 1918), although it was in a less outspoken political perspective. They both focused on learning through experience and action. Interaction and problem solving were core elements in this approach and Kilpatrick developed the idea of educational projects where students engage in purposeful activities. Teachers were given a new role by guiding these learning processes.

These ideas of learning did exist at the time of establishing the reform universities; however, the connection between Dewey, Kilpatrick, or Freire and the establishment of new pedagogy and learning methodologies is more an after-reflection than a guide for designing new educational approaches.

In engineering education in particular, companies require candidates with a high level of knowledge and skills who are able to participate in innovation projects and understand collaboration (National Academy of Engineering, 2004; Royal Academy of Engineering, 2007). This requirement has been formulated since the 1990s and before, e.g. Bucciarelli has pointed out that woven into technological knowledge was a constant process of negotiation and collaborative understanding of the design process (Bucciarelli, 1994).

With a global technology development, the ability to handle complexity and intercultural collaboration is an embedded engineering capacity, and innovation is based on collaborative knowledge construction (Sawyer & Sawyer, 2005). This need does not decrease and the requirement for complex competencies will grow with the need for more 'green' engineers (Jamison, 2013).

The fact that PBL was an institutionalized educational practice has definitely contributed to the success and dissemination of this pedagogy. The institutional part involves organizational structures built into the curriculum, and the practice did not just cease if one academic staff member moved from the institution. It was possible to create a culture and build up a sustainable community through new recruitment and educational training. Even if there was skepticism among the academic staff, they became more optimistic when the first reports on graduates were positive. During the 1980s, the reform universities proved their value by receiving very good responses from the companies that employed the graduates.

3.5 Comparing PBL Universities

Ginie Servant has compared the four main reform and PBL-like universities as seen in Table 3.2. Based on interviews with some of the pioneers, scholarly literature, and documents from that time, she has not only written the history but also pointed to the fact that there are variations in both the historic reasons for founding these new universities and the variations in the pedagogic and curriculum approach.

As Table 3.2 indicates, there are quite some differences concerning the educational format, the role of the teachers, the problems, types of problems, cycle of the problems, team sizes, relations between lectures and students' work, disciplines in which this unfolded, and many other variables. The variation would even be larger if the analyses had moved from an institutional level to a faculty or department level, as we get closer to practice.

What is notable from Table 3.2, along with the analyses by Ginie Servant is that engineering disciplines are applying PBL only at Aalborg University. McMaster and Maastricht are known for medicine and health sciences, and there are also some social science disciplines at Maastricht. Roskilde and Aalborg have a broader range of disciplines, although engineering is only at Aalborg. The research on PBL grew out of the engineering faculty by the establishment of first the UNESCO Chair in PBL and, later, the UNESCO category 2 Centre on PBL in Engineering, Science and Sustainability.

During the last 30–50 years there has been a global change mode with external requests to learn about projects in engineering. Technology is a result of collaborative creation and production processes and is in its nature project-based, as innovation is about developing new products and applications. So, there has been a clear match between engineering disciplines and the needs of technological development and project work.

The importance of social concerns seems to be less important in engineering and science compared to social science and humanities. The understanding of problems varies according to the disciplines. It does not mean that problems in engineering are purely technical, as technology has a huge impact on society and humans.

There are other institutions which have developed systemic PBL practices rather different from the universities mentioned in Table 3.2. University of Delaware, USA, and Republic Polytechnic in Singapore are just two of the institutions which have developed different versions of PBL. University of Delaware has a PBL program adjusted to the various courses (Savery, 2015). Republic Polytechnic has an

Table 3.2 Comparing the four reform universities (Servant, 2016, p. 237)

	McMaster	Maastricht	Roskilde	Aalborg
Year of entry of first cohort	1969	1974	1972	1974
Historical reasons for reform	Unhappy with old ways	Political decisions to justify building the eighth medical faculty	Defy traditional authority, social revolutionary considerations, some pedagogical considerations	Defy traditional authority, improve engineering education
Role of teachers	Tutors no longer lectured, process-focused, but still acted as group leaders.	Tutors no longer lectured, process-focused, did not have to be content experts, students were group leaders.	In projects (50%) Teachers are seen as equal members of the group, as methods & content resource persons for students. In courses (50%) Lecturer	In courses (50%) In principle the same as Roskilde but some tension with teachers coming in from traditional institutions. In courses (50%) Lecturer
Education format	Mainly: small-group tutorial. Also: Guided instruction, developmental discussion, field trip, self-learning, lecture, recitation.	Small-group tutorials *Skills lab* Field work	1972–75: 100% project-work in Soc Sci. & Hum. 50% project v. 50% courses in Nat. Sci After 1975: 50% project work 50% courses	Before 2010: 50% project work 25% project courses 25% courses After 2010: 50% project work 50% courses
Group size	Smaller groups	Small groups	Small groups 50% Large groups 50%	Small groups 50% Large groups 50%
Origin of problems	Given by faculty	Given by faculty	Formulated by students based on themes	Formulated by students based on themes
Problem structure	1969–1977: Biomedical problems 1977–1993: Socially relevant medical cases 1993-Present: Biomedical problem	Biomedical problems Social problems Medical cases	Social problems Real-life problems	Hum & Soc Sci: Social problems real-life problems Engineering and science: Real-life problems, technical problems
Problem cycle	Short	Short	Long	Long

(continued)

Table 3.2 (continued)

	McMaster	Maastricht	Roskilde	Aalborg
Problem deliverables	None	Group problem report (ungraded)	Group problem report (graded)	Group problem report (graded)
Interdisciplinarity	Some attempt to do away with traditional medical disciplinary divisions.	Some attempt to do away with traditional medical disciplinary divisions, but neither explicit nor important.	Central to the model. Attempt to go beyond the disciplines failed in the late 1970s. Model changed to 'multidisciplinary' in 1980s.	Initially central to the model, but progressive return to disciplines in the 1970s.
Importance of social concerns	Some, but never really played out in the curriculum.	Some, initially important for founding Dean but less prominent later.	Central to the model	Central to the model in social sciences, less in technical & natural sciences.
Importance of training professional skills	Somewhat important, although 'skills' training didn't come in until the late 1970s.	Very important, invention of the *Skills lab.*	Not important, technical sciences deliberately excluded.	Hum & Soc Sci Not very important Tek-Nat very important
Importance of political concerns	None	None	Central to the model	Some in the beginning, quickly fading
Assessment	Initially only peer-evaluation. Then introduction of 'Triple Jump'. 1993 Introduction of progress test	Block tests Progress Test (individual)	Group exams based on project group Initially no course exams, later individual exams for courses	Group exams based on project group Individual exams for courses
Disciplines of model application	Medicine & Health Sciences	Medicine, Health Sciences, Law, Economics, Psychology, etc.	Humanities, Social Sciences, Natural Sciences	Humanities, Social Sciences, Technical & Natural Sciences

institutional profile with a 'one day, one problem', where the students work in team on teacher defined problems which can be analyzed and solved within one day (O'Grady et al., 2012).

Among the various intuitional PBL strategies, there are core differences in the organization of the curriculum, the teachers' role, group size, the way the PBL activity is organized, the identification of problems, team processes, and the assessment as well if PBL is also embracing team projects or cases. PBL can be very limited with a one day one problem approach or can be broader like in the social sciences at Aalborg University. In my view, this is not at all problematic and in Chap. 5, I will sum this up with some PBL learning principles. I find it enlightening that there are very many variations, and in this way, there will always be inspiration to be found from real practices.

However, the issue is that for most of the institutions mentioned, the single institution maintains a PBL practice within a single band, and therefore the students will only experience one type of PBL. What is even more worrying is that it seems rare that the students can propose the problems with only Aalborg University seemingly providing this opportunity. For the McMaster system, the students start out with cases, which can be more openly formulated or narrower and therefore it will depend on the character of the problem.

In a project-based system, the openness and closedness also count and it is a question of curriculum practices. Open or ill-structured problems allow students to create ownership during the process of analysis and decision making. Ownership is core in motivation for learning and trying out new things.

Routines and routine learning are traps also existing in PBL systems. If you are used to the seven jumps method, you might find it boring after using it more than three times. When you do projects worth 15 ECTS, this might appear boring as well as you already know how to start up the project, how to share the work, end the project, and how to defend it at the project exam. You do learn something new in relation to the problems and the scientific content, but you learn limited new skills and competencies from the project process. The same counts for the one day one problem approach as the learning process is organized in the same way.

However, there are also institutional changes. For example, there has been a move from interdisciplinary to disciplinary approaches, and the pendulum has yet to swing back (Neville & Norman, 2007). Interdisciplinary or disciplinary approaches influence the PBL learning process, notably in terms of the students' ability to select problems to work on in their projects.

3.6 Other Taxonomies for PBL

There are other taxonomies for PBL pedagogies. Barrows developed a taxonomy identifying different levels of student-centered learning. He wrote:

> The increasingly popular term 'problem-based learning' does not refer to a specific educational method. It can have many different meanings depending on the design of the educational method employed and the skills of the teacher. (Barrows, 1986, p. 481)

He developed the following taxonomy for categorizing PBL practice in medicine:

- Lecture-based cases: Cases used to demonstrate the relevance of information provided by lectures.
- Case-based lectures: Cases are used to highlight material that will be covered in the subsequent lecture.
- Case method: Cases are studied in preparation for class discussion.
- Modified case-based method: Cases provide opportunities for deciding between a limited number of options.
- Problem-based learning: Cases are used in a problem simulation format that encourages free inquiry.
- Closed loop, or reiterative, problem-based learning: A reflective phase complements the problem-based format.

This taxonomy range demonstrates how cases can illustrate points at one narrow end and a more open and reflective PBL approach on the other, resulting in a question of instructor control versus student control. However, it is vital to note that this taxonomy only addresses the course/unit level, not the program/educational/institutional level. It is a taxonomy created for studying and developing cases for classroom teaching. It does demonstrate the variety of cases that can be used, as well as the differences in how these cases are employed as part of the students' learning process and the lecturer's teaching process. There is a significant distinction between using cases as demonstrations in lecturing and allowing students to design their own problems in teamwork.

Savin-Baden proposes different modes of PBL covering the objective of PBL which include the perception of knowledge, learning, problems, students, teacher roles, and assessment (Savin-Baden, 2000) (see Table 3.3). These modes hypothesize five different aligned models of PBL practice covering the intended relation among learning, problem scenario, students, facilitators, and assessment, although the starting point is clearly the epistemology of the problem.

In contrast to Barrows' approach (1986), Savin-Baden's models are more comprehensive and align knowledge with types of problems. She emphasizes that open-ended problems cannot effectively address knowledge objectives, such as propositional knowledge. Learning objectives must align with appropriate problem types, learning processes, facilitator roles, and assessments. The six dimensions across the five models emphasize that implementing PBL involves not merely a shift in learning methodology, but rather an integration of learning methodology, knowledge construction, and a scientific approach.

Consequently, transitioning to a full-scale PBL model influences the scientific approach by training students to employ research methodologies and critically evaluate the propositional knowledge obtained from academic sources. Analyzing and solving problems requires a pragmatic approach to the interplay between theory and practice, alongside the advancement of theoretical and analytical comprehension across established knowledge boundaries.

Table 3.3 Models of problem-based learning (Savin-Baden, 2000)

Model I *Problem-based learning for epistemological competence*	Knowledge: Propositional knowledge Learning: Use of propositional knowledge Problem scenario: Limited problems with known solutions Students: Are receivers of propositional knowledge Facilitators: Guide to understand the correct knowledge Assessment: Test of knowledge according to objectives
Model II *Problem-based learning for professional action*	Knowledge: Know-how Learning: Skills for workplace Problem scenario: Real life Students: Learning to solve real problems in order to undertake practical action Facilitators: Demonstrator of practical skills Assessment: Test of skills for the workplace and supporting knowledge
Model III *Problem-based learning for interdisciplinary understanding*	Knowledge: A cross between know-how and know-what Learning: Knowledge and skills across discipline boundaries Problem scenario: Centered around knowledge with action Students: Integrators across boundaries Facilitators: Coordinator of skills and knowledge across boundaries Assessment: Skills and knowledge in a relevant context
Model IV *Problem-based learning for transdisciplinary learning*	Knowledge: To understand existing boundaries Learning: Critical thinking and understanding discipline borders Problem scenario: Dilemmas Students: Independent thinkers with a critical stance Facilitators: Flexible Assessment: Opportunity to demonstrate integration across disciplines
Model V *Problem-based learning for critical contestability*	Knowledge: Contingent, contextual and constructed Learning: Interrogation of frameworks Problem scenario: Multidimensional, offering alternative ways of understanding Students: Explorers of underlying structures and beliefs and developing new hypotheses and knowledge Facilitators: Commentator, challenger and decoder of cultures, disciplines and traditions Assessment: Open-ended and flexible

3.7 The Importance of the Problem

As previously described, there are variations among the institutional models, and there are also variations in the core elements of PBL: the problem and the project, and how each is used in the curriculum. As Savin-Baden has indicated, there are many different approaches to problems. The Danish traditions contribute with focus on how students can identity and solve problems. The discussion on what is a problem characterizes one part of the literature.

As already indicated, student-centeredness and ownership of the problem characterize the Aalborg model. From the origins of the Danish traditions, Hultengren sees a problem as a wonder, something one would like to know more about and experience, or something that you find is scandalous (Hultengren, 1981). Algreen-Ussing and Fruensgaard divide the problem into phases and formulate the first phase as the initiating problem, followed by a problem analysis phase, and formulation of the specific problem to be solved. The initiating problem is the trigger for motivation; however, it has to be analyzed and defined before it can be applicable in an engineering project, and it is the students with an advisor who are the core actors in this process (Algreen-Ussing & Fruensgaard, 1992).

My colleagues, Helle Algreen Ussing and Niels Fruensgaard, who introduced me to PBL in engineering education, have done significant work to develop a PBL language and methods. The most well-known are the nine project phases, which began with problem identification and analysis. These first two phases include the relation to society, where the initiating problem was the trigger, something of interest or curiosity.

During the first-year program it could be, for example, that some of the students knew friends working in kindergartens where the noise level had increased after the removal of asbestos ceilings. The students went out to analyze the building and materials and measure the noise level at different times during the day. After the problem analysis, there are phases to limit the problem and formulate requirements. This problem can lead to projects on architecture for architecture and design students, projects on materials for noise and vibration reduction in buildings, civil engineering projects on building construction, or electronics projects on measurement. The final phases involved solving the problem, developing implementation strategies, and evaluating them (see Table 3.4).

In comparing these phases to the Maastricht model, we see there are visible differences. The purpose of the problem and the problem analyses are also different in comparison to the Maastricht model. Erik de Graaff and I compared two types of problem analysis guides: the seven jumps and the project phases. The purpose of case-based PBL models is to understand the problem, whereas the project phases outlines for engineering and science students are to solve a problem based on a shorter phase of identification.

These two approaches to problem analyses reflect the disciplines for which they were developed. In both the social sciences and natural sciences, the understanding

Table 3.4 Differences between the seven jumps method and project phases (Anette Kolmos & Erik de Graaff, 2014, p. 150)

	Seven jumps (Wilkerson & Gijselaers, 1996)	Project phases (Algreen-Ussing & Fruensgaard, 1992)
Problem analysis	1. Clarify terms and concepts not readily comprehensible 2. Define the problem 3. Analyze the problem and offer tentative explanations 4. Draw up an inventory of explanations 5. Formulate learning objectives 6. Collect further information through private study 7. Synthesize the new information and evaluate and test it against the original problem. Reflect on and consolidate learning.	1. Initiating a problem (the trigger for the problem—what starts it) 2. Problem analysis (analysis of the problem—for whom, what and why) 3. Definition and formulation of problem (specification requirement)
Problem solving		4. Problem-solving methodologies (overview of possible solutions and assessment of impact) 5. Demarcation (argumentation for the choice of solution) 6. Specification of requirements 7. Solving the problem (carry out the solution—construction/design/further analysis) 8. Implementation (prototype and sometimes real systems) 9. Evaluation and reflection (impact, effect and efficiency of solution).

of theories and problems are core learning outcomes, whereas the learning outcomes in engineering include both an understanding of the problem and an understanding of how to solve it.

3.8 Variation in the Problem Approach

In the book *The Aalborg PBL model*, Palle Qvist analyzes the literature at that time and finds various understandings of the problem as the trigger for the learning: conflict, wonder, something that has or is or will happen, knowledge about something experienced, scandal, curiosity, lack of knowledge, lack of function, uncovered need, wish, practical problem, contrast between a wish and reality, or, in more general terms, an anomaly, a paradox, contrast, or a contradiction (Qvist, 2004). This indicates a broad understanding of the problem.

From recent studies, the problem is seen as the core anchor during the whole process. The problem is formulated and re-formulated many times, and a result was found by one of the PhD students who following a project team (Velmurugan, 2022) who realized that the problem was the anchor for the whole project from the beginning to the end, and that there were several re-formulations of the problem during that process. The fact that the students have the responsibility of identifying problems and carrying out the argumentation for the problem is a huge part of the collaborative process in the projects.

In case-based PBL literature, the main characteristic is to conceptualize design processes for academic staff as they are responsible for the problem identification. Even if there can be broad or narrow problem statements in the cases, it is the responsibility of academic staff to formulate it. Schmidt and Moust approach problems with the aim of explaining the kind of knowledge the students would work to achieve (Smith et al., 2022). They distinguish between:

- explanatory knowledge (theories) represented by explanatory problems
- descriptive knowledge (facts) by fact-finding problems
- procedural knowledge by strategy problems
- subjective knowledge by moral and dilemmas (Schmidt & Moust, 2000a).

A recent study underlines that PBL needs a rich problem that is authentic, open ended, and complex. It is not possible to address complex problems without referring to the work by Jonassen, who has done incredible work on problem-solving competencies (Jonassen, 2010). Problem-solving is very often used as the term in the US-engineering area, where there is a phase of problem identification in the problem-solving process. Translating the problem in PBL into problem-solving has caused criticism from PBL researchers in engineering and especially the humanities and social sciences. Within engineering disciplines, it has been argued that for students, problem analysis is important to open opportunities in innovation processes, and in the social sciences and humanities the problem analysis phase is the project.

Jonassen has listed an overview of 11 different types of problems: logic problems; algorithms; story problems; rule-using/rule-induction; decision making; troubleshooting; diagnosis-solution problems; strategic performance; policy-analysis problems; design problems; dilemmas (Jonassen, 2010, p. 11).

For each problem type, there are external criteria that create variations (Jonassen, 2010, p. 6):

1. Structuredness with a continuum between ill-structured and well-structured problems where it is mostly well-structured problems that are applied in formal education.
2. Context of the problem also relates to structuredness, as everyday problems or authentic problems, which are normally ill-structured.
3. Complexity is a question of the number of interacting factors. Ill-structured everyday problems normally have a higher degree of complexity compared to, for example, textbook math.
4. Dynamicity is a part of complexity as the interaction among factors change over time, and therefore the complexity analyzed at an early stage might have changed when the solution is chosen and developed.

5. Domain specificity concerns the different cognitive strategies or logics for cognitive domains, for example, psychology versus engineering, where the argumentation for problems will take very different starting points.

Jonassen's handbook is not written for the PBL community, and there are no explicit links to the PBL communities in how to apply these variations of problems in design processes, formulation of learning outcomes or guidelines for students. But it is a great logical conceptualization of the variations in problems and a way to analyze the complexity of problems. Most often, only structuredness and complexity are mentioned as dimensions for problem analyses; however, the three other dimensions give a deeper understanding, e.g. complexity is not only the number of interacting factors but also time, actors' perspectives, and the scientific approach.

The Cynefin framework, as described in Chap. 2, is often used to describe characteristics of problems, ranging from narrow, complicated, complex and chaotic (Habbal et al., 2024; Hadgraft & Kolmos, 2020). Complex problems have been on the agenda in engineering education for quite some time. Complex problems are characterized by the number of relations among factors, and the issue is that when bringing them into education, we will have cut a section out and reduce the number of interfering variables and reduce the complexity.

We will have the same situation if we are going to work on chaotic situations in engineering education. In chaos, there are unpredictable interrelations and variables beyond our control that we need to learn how to deal with. As already mentioned, we need to bring in elements of this chaos in projects or cases where students can learn to act along with sensing.

3.9 Reflection 3: Living-With Chaos

I have tried to indicate the variations of PBL, which can give inspiration for development of learning. I also wanted to emphasize that the PBL we know today has its origins in broader values. There was an intention of educating critical citizens who could be key figures in existing society and create a better place to live as well as integrate societal issues and problems. The original PBL models were born with an ideology and imagination beyond the knowledge components. Maybe we cannot apply this ideology and critical stance today; however, we can reformulate it to address the complex problems and chaos that society witnesses. We can start to talk about chaos and how to handle chaotic problems.

Knowledge is a fundamental response to understanding societal problems and responding to them; however, this is no longer enough as we are dealing with young people's anxiety, which is felt at a personal level, and when we are in chaotic situations, we need to react immediately.

The fact that this pedagogy opens up an academic understanding of societal problems can be a very important element in learning to live with chaos. The interaction between academic and societal practice can take different forms and chaos in society or private life is not something that we usually speak about in education. We have started to bring in complexity to education although this is hardly chaos. Such knowledge can create preparedness at a rational level, especially if students have learned to approach problems of different formats and can solve them in various collaborative formations. This overview of PBL institutional models and possible variations in problems might therefore serve as inspiration for curriculum development. We do not need to invent everything from scratch; it is already on the other side of the fence.

Alternative models are great for visioning new systems with different types of alignments among curriculum elements. Although when institutional PBL models are established, these are as hard to change again as with all other types of educational practice. For example, in medicine there are written tutorials for cases in medicine to guide the tutors, and if the students get the tutorial somehow, they know the expected results and are able to see any lack of learning.

We do not need to write new textbooks and therefore the openness in the problem and the fact that students define and own the problems are crucial for learning to foster resilience. Maybe there should be progression for students to move from something more teacher-directed to student-directed; however, the goal should be that students learn to be independent drivers of their own individual and collaborative learning processes.

What is also important to learn from this chapter is that pedagogical ideas have a very different understanding of learning as experience-based and problem-oriented learning, where empowerment and emancipation were in the core understanding of the pedagogy. Perhaps it was another type of context in which the students had to liberate and free themselves from societal constraints; however, it was a discourse during which emancipation and empowerment were discussed.

For emancipation, today there is a new understanding in the direction of 'woke' and liberation for minorities of both race and gender, which include political correctness. This is a totally different discourse to the battle during the 1970s concerned the capitalistic system, as the woke debate affects cultural and behavioral issues. No matter the character of the battle, there is a call for empowerment in relation to resilience.

References

Algreen-Ussing, H., & Fruensgaard, N. O. (1992). *Metode i projektarbejdet: problemorientering og gruppearbejde* (3. opl. ed.). Aalborg Universitetsforlag.

Baillie, C., & Kadetz, P. I. (2024). *Reimagining engineering education*. Springer.

Barrows, H. S. (1986). A taxonomy of problem-based learning methods. *Medical Education, 20*(6), 481–486.

Barrows, H. S. (1996). Problem-based learning in medicine and beyond: A brief overview. *New Directions for Teaching and Learning, 1996*, 3–12. https://doi.org/10.1002/tl.37219966804

Berthelsen, J., Illeris, K., & Poulsen, S. C. (1977). *Projektarbejde*. Borgen.

Beyer, L. E. (1997). William Heard Kilpatrick (1871–1965). *PROSPECTS: the quarterly review of comparative education* (Paris, UNESCO: International Bureau of Education), vol. XXVII, no. 3, 470–85.

Bitsch Olsen, P., & Pedersen, K. (2005). *Problem-oriented project work: A workbook* (1st ed.). Roskilde University Press.

Bucciarelli, L. L. (1994). *Designing engineers*. MIT Press.

Chen, J., Kolmos, A., & Du, X. (2020). Forms of implementation and challenges of PBL in engineering education: A review of literature. *European Journal of Engineering Education*, 1–26.

Crawley, E. F., Malmqvist, J., Östlund, S., Brodeur, D. R., & Edström, K. (2014). The CDIO approach. In *Rethinking engineering education* (pp. 11–45). Springer.

De Campos, L. C., Dirani, E. A. T., & Manrique, A. L. (2012). *Project approaches to learning in engineering education: The practice of teamwork*. Springer.

Dewey, J. (1903). Democracy in education. *The Elementary School Teacher, 4*(4), 193–204.

Dewey, J. (1958). *Experience and nature* (Vol. 471). Courier Corporation.

Dikilitaş, K., Marshall, T., & Shahverdi, M. (2025). *A practical guide to understanding and implementing challenge-based learning*. Palgrave Macmillan.

Drake, S., & Burns, R. (2004). *Meeting standards through integrated curriculum*. ASCD.

Felder, R. M., Brent, R., & Prince, M. J. (2015). Engineering instructional development: Programs, best practices, and recommendations. In *Cambridge handbook of engineering education research* (pp. 409–436). Cambridge University Press. https://doi.org/10.1017/CBO9781139013451.027

Freire, P. (1973). *Education for critical consciousness*. Bloomsbury Publishing Kindle edition.

Graham, R. (2010). *UK approaches to engineering project-based learning. White paper sponsored by the Bernard M. Gordon-MIT Engineering Leadership Program*. https://www.rhgraham.org/resources/MIT-White-Paper---UK-PjBL-April-2010.pdf

Habbal, F., Kolmos, A., Hadgraft, R. G., Holgaard, J. E., & Reda, K. (2024). *Reshaping engineering education: Addressing complex human challenges*. Springer Nature.

Hadgraft, R. G., & Kolmos, A. (2020). Emerging learning environments in engineering education. *Australasian Journal of Engineering Education, 25*(1), 3–16.

Hultengren, E. (1981). *Problemorientering, projektarbejde og rapportskrivning*. Aalborg Universitetscenter, Institut for uddannelse og socialisering.

Hunsu, N. J., Carnell, P. H., & Sochacka, N. W. (2021). Resilience theory and research in engineering education: What good can it do? *European Journal of Engineering Education, 46*(6), 1026–1042.

Illeris, K. (1974). *Problemorientering og deltagerstyring: oplæg til alternativ didatik*. Munksg@ård.

Jamison, A. (2013). The making of green engineers: Sustainable development and the hybrid imagination. *Synthesis Lectures on Engineering, 8*, 1–153. https://doi.org/10.2200/S00476ED1V01Y201301ENG020

Jonassen, D. H. (2010). *Learning to solve problems: A handbook for designing problem-solving learning environments*. Routledge.

Kilpatrick, W. H. (1918). *The project method: The use of the purposeful act in the educative process*. Teachers College, Columbia University.

Kjærsdam, F., & Enemark, S. (1994). *The Aalborg experiment project innovation in university education*. Aalborg Universitetsforlag.

Klafki, W. (2007). *Neue studien zur bildungstheorie und didaktik*. Beltz.

Kluge, A., & Negt, O. (2016). *Public sphere and experience: Analysis of the bourgeois and proletarian public sphere*. Verso Books.

Kolmos, A. (2017). PBL curriculum strategies: From course based PBL to a systemic PBL approach. In *PBL in engineering education: International perspectives on curriculum change* (pp. 1–12). Sense Publishers. https://doi.org/10.1007/978-94-6300-905-8

Kolmos, A., & de Graaff, E. (2014). Problem-based and project-based learning in engineering education. In B. M. Olds & A. Johri (Eds.), *Cambridge handbook of engineering education research* (pp. 141–161). Cambridge University Press.

Kolmos, A., Fink, F., & Krogh, L. (2004a). The Aalborg model. In A. Kolmos, F. Flemming, & K. Lone (Eds.), *The Aalborg model: Progress, diversity and challenges* (pp. 9–18). Aalborg Universitetsforlag.

Kolmos, A., Fink, F., & Krogh, L. (2004b). *The Aalborg PBL model: Progress, diversity & challenges*. Aalborg University Press.

Kolmos, A., Holgaard, J. E., & Clausen, N. R. (2021). Progression of student self-assessed learning outcomes in systemic PBL. *European Journal of Engineering Education, 46*(1), 67–89.

Korsgaard, M. T. (2019). Exploring the role of exemplarity in education: Two dimensions of the teacher's task. *Ethics and Education, 14*(3), 271–284.

Lehmann, M., Christensen, P., Du, X., & Thrane, M. (2008). Problem-oriented and project-based learning (POPBL) as an innovative learning strategy for sustainable development in engineering education. *European Journal of Engineering Education, 33*(3), 283–295. https://doi.org/10.1080/03043790802088566

McEwen, K. (2011). *Building resilience at work*. Australian Academic Press.

Moesby, E. (2008). *What is an effective approach to introducing PBL/POPBL in an institution*. Institut for Samfundsudvikling og Planlægning, Aalborg Universitet.

National Academy of Engineering. (2004). *The engineer of 2020: Visions of engineering in the new century*. National Academy of Engineering. Retrieved January 17, 2007, from http://books.nap.edu/catalog/10999.html

Negt, O. (1971). *Soziologische Phantasie und exemplarisches Lernen [Sociological imagination and exemplary learning]*. EuropaÈische Verlagsanstalt.

Negt, O. (1974). *Soziologische Phantasie und exemplarisches Lernen: zur Theorie d.* Europäische Verlagsanstalt.

Neufeld, V., Norman, G., Feightner, J., & Barrows, H. (1981). Clinical problem-solving by medical students: A cross-sectional and longitudinal analysis. *Medical Education, 15*(5), 315–322.

Neville, A. J., & Norman, G. R. (2007). PBL in the undergraduate MD program at McMaster University: Three iterations in three decades. *Academic Medicine, 82*(4), 370–374.

Neville, A., Norman, G., & White, R. (2019). McMaster at 50: lessons learned from five decades of PBL. *Advances in Health Sciences Education, 24*, 853–863.

Norman, G. (1998). PBL - The least worst curriculum design? *Advances in Health Sciences Education, 3*(1), 1–2.

O'Grady, G., Yew, E., Goh, K. P., & Schmidt, H. (2012). *One-day, one-problem: An approach to problem-based learning*. Springer.

Pblworks. (2025). https://www.pblworks.org/what-is-pbl

Pellew, J., & Taylor, M. (2020). *Utopian Universities*. Bloomsbury Publishing.

Prince, M. J., & Felder, R. M. (2006). Inductive teaching and learning methods: Definitions, comparisons, and research bases. *Journal of Engineering Education-Washington, 95*(2), 123.

Qvist, P. (2004). Defining the problem in problem-based learning. In A. Kolmos, F. Fink, & L. Krogh (Eds.), *The Aalborg PBL model progress, diversity and challenges*. Aalborg Universitetsforlag.

Royal Academy of Engineering. (2007). *Educating engineers for the 21st century*. http://www.raeng.org.uk/publications/reports?p=6

Rugarcia, A., Felder, R. M., Woods, D. R., & Stice, J. E. (2000). The future of engineering education I. A vision for a new century. *Chemical Engineering Education, 34*(1), 16–25.

Savery, J. R. (2015). Overview of problem-based learning: Definitions and distinctions. In *Essential readings in problem-based learning: Exploring and extending the legacy of Howard S. Barrows* (Vol. 9, pp. 5–15). Purdue University Press.

Savin-Baden, M. (2000). *Problem-based learning in higher education: Untold stories*. Society for Research into Higher Education, Open University Press.

Sawyer, R. K., & Sawyer, R. K. S. (2005). *Social emergence: Societies as complex systems*. Cambridge University Press.

Schmidt, H. G., & Moust, J. H. (2000a). Towards a taxonomy of problems used in problem-based learning curricula. *Journal on Excellence in College Teaching, 11*(2&3).

Schmidt, H. G., & Moust, J. H. C. (2000b). Factors affecting small-group tutorial learning: A review of research. In *Problem-based learning: A research perspective on learning interactions* (pp. 19–52). Academia.

Servant, V. (2016). *Revolutions & re-iterations: An intellectual history of problem-based learning*. Erasmus University College (EUC).

Servant, G. (2024). *Pedagogies of collapse: A hopeful education for the end of the world as we know it*. Bloomsbury.

Servant-Miklos, V. F., Norman, G. R., & Schmidt, H. G. (2019). A short intellectual history of problem-based learning. In *The Wiley handbook of problem-based learning* (pp. 3–24). Wiley.

Smith, K., Maynard, N., Berry, A., Stephenson, T., Spiteri, T., Corrigan, D., Mansfield, J., Ellerton, P., & Smith, T. (2022). Principles of problem-based learning (PBL) in STEM education: Using expert wisdom and research to frame educational practice. *Education Sciences, 12*(10), 728.

Velmurugan, G. K. (2022). *Problem construction in problem-based learning: How students deal with disagreements in decision-making*. Aalborg Universitetsforlag.

Wagenschein, M., & Horton-Krüger, G. (2012). Teaching to understand: On the concept of the exemplary in teaching. In *Teaching as a reflective practice* (pp. 161–175). Routledge.

Wiggershaus, R. (1994). *The Frankfurt School: Its history, theories, and political significance*. MIT Press.

Wilkerson, L., & Gijselaers, W. H. (1996). *Bringing problem-based learning to higher education: Theory and practice*. Jossey-Bass.

Wright, S. (2025). Changes in university 'autonomy' with successive visions of Denmark's future. *Higher Education, 89*(2), 331–348.

Yew, E. H., & Schmidt, H. G. (2009). Evidence for constructive, self-regulatory, and collaborative processes in problem-based learning. *Advances in Health Sciences Education, 14*, 251–273.

Chapter 4
A Case of Aalborg University: The Pioneers and Critical Voices

4.1 Introduction

I have in the previous chapter described some of the theoretical ideas and intentions behind the PBL pedagogy. Even though there have been several theoretical considerations behind the establishment of PBL models at McMaster, Linköping, Maastricht, Roskilde, and Aalborg Universities, the development of Aalborg University engineering education throughout the 1970s and 1980s has been characterized by pragmatism, and driven by the question: What works in practice? The teachers at the institutions developed their practices and if something didn't work well, they tried something else. It was an innovative, creative, experiential, and experimental process that can exist well in smaller organizations in the pioneering phase.

As most of my empirical work and playground for development has been Aalborg University, I will go deeper into the PBL history with the development of the Aalborg PBL model. This has been described in many different publications, and my intention is not to repeat it, but I do want to tell a story of a pioneer culture that belongs to a certain phase of the establishment of new alternative pedagogics.

Nothing is created without enthusiasm, energy, collaboration, and vision. However, the ideas and academics who committed themselves to this work often vanish from history as they did not write about it. I call for a new wave of educational transformers and there must be something we can learn from the significant changes made more than 50 years ago.

Aalborg University, located in the north part of Jutland, was established in 1974. During the 1970s, it took seven hours by train to get from Aalborg to Copenhagen; there was no motorway and even if there had been a well-established airport, it was not common at the time to go by plane. The infrastructure has developed significantly since then and Aalborg today is no longer an isolated island. This is

A. Kolmos, *Engineering Education for Interdisciplinary Futures*, Emergent Directions in Engineering Education 2,
https://doi.org/10.1007/978-3-032-32540-2_4

mentioned as the university was physically situated in a significantly non-academic environment, far away from Copenhagen.

During the 1970s, several new universities were established as universities moved from an elite to a more mass-market approach. This also meant there was a large group of first-generation students coming from non-academic backgrounds. New universities without long academic traditions and with new first-generation academics provided a special environment for innovation. What most academics and students experienced during those first decades was a pioneering spirit and a feeling that everything was possible—within limits.

Many academics and students dedicated their entire lives to the establishment of this new university, not to mention the development of its pedagogy. At that time, we did not know the procedures and the academic cultures and were building the culture as we were building up the university. This pioneering spirit somehow bound us together and we could play, experiment, reflect, and reform the education. I am and was a part of this group and am now on my way to disembarking the ship.

Many first-generation students had a working-class background or came from self-employed farmers and merchants. Children with these social backgrounds had generally already learned to work; during childhood they were given responsibilities and tasks to do. That was part of being brought up during the 1960s. Somehow this aligned very well with the project work as it requires a certain social discipline and work ethic.

The foundation of Roskilde University in 1972 and Aalborg University in 1974 were the result partly of an expanding economy with the need for more academics and the student movement, which had required reforms of the traditional universities (see Chap. 3). The Danish left-wing students found inspiration from Germany and France at that time and the critical theory of Negt, Kluge and Habermas along with the logic of capital by Marx were the foundations for the new pedagogy—or so we thought (see Chap. 3).

The North part of Jutland had been petitioning for a university for a long time, and finally one of the new Danish universities would be located in that part of the country. It was also decided that the two new universities of Roskilde and Aalborg should be established with an alternative pedagogical model based on problem-orientation and collaborative project work, which was pushed from a strong student movement.

Aalborg University started out by merging two engineering schools with a core focus on engineering education and the technical sciences. Engineering builds systems and involves project collaboration and in that sense the project work with real-life problems should be a logical choice for engineering.

The pioneer spirit is related to small entities that have a common goal. The pioneers are the first on new ground and have a willingness to work hard to live in new places or try out new things. Building a university is not necessarily to conquer new land, but to explore and investigate. Aalborg University began in the slipstream of Roskilde University, which took on many of the political battles as it was located much closer to Copenhagen and was visible on the political scene. The fact that engineering dominated at Aalborg University was also one of the reasons that it went under the radar from the politics. In engineering, the political agenda has never dominated the discourses as compared to social science and the humanities.

4.2 Skepticism of Aalborg University and the Renaming to PBL

Resistance against the change of the teaching and learning paradigm was, in the beginning, significant among the academic staff embedded in the merger process when the two engineering schools changed the teaching and learning paradigm; however, resistance from the professional engineering society was even worse. In the beginning, the Danish Association for Engineers, which at that time was dominated by engineers educated from the Technical University, would not accept engineers who had graduated from Aalborg University. In addition, the industry in general had to adjust and believe in the competencies of engineers graduating from this new problem-oriented and project-organized engineering curriculum. These companies, however, soon responded positively to the new competency profiles (Jensen & Wagner, 1990).

The most explicit skepticism came from academia itself. When you have earned your graduation and gained your professional identity in one learning regime, it might be hard to trust new ways of teaching and learning. Culture repeats itself, and learning creates inner values and identity structures, which is one of the reasons that it will always be hard to transform an existing system. During the 1980s and 1990s, there were tensions between Aalborg University and the well-established Danish universities; thankfully, this is no longer the case.

In my first international journal article, I compared problem-based learning and project work and found many similarities and differences (Kolmos, 1996). This article has been cited as being the changing point for Aalborg University moving from calling the system problem-oriented project work to PBL (Servant-Miklos, 2020). This is not correct as at that time I did not call the project work PBL; I was just comparing the Aalborg and Maastricht systems.

The renaming of the pedagogical model from problem-oriented project work to PBL came later and was decided at the institutional level. This was after a period establishing UNISCENE (the University Student Centered Network), which was a group of the new reform universities exchanging experiences. The UNISCENE network consisted of Maastricht, Roskilde, Linköping, Tromsø, and Aalborg Universities, all which were newly established and attempting to develop and practice new pedagogies. There were several anniversary conferences as Maastricht and Aalborg both started to look for more theoretical reasons for the success of their PBL models.

As already described, the Maastricht and the Aalborg models were different on significant points, as Maastricht was working with teachers who formulated cases of six weeks, whereas Aalborg was semester projects with freedom for students to identify their problems within a theme. However, the understanding of the learning theories behind and the focus on student-centeredness were the same. Later my colleague Erik de Graaff and I wrote an article on problem-based and project-based learning merging models (De Graaff & Kolmos, 2003). The case-based model, which worked very well in medicine, did not do so well in engineering and vice versa due to disciplinary differences.

The Aalborg model was renamed as a PBL model, and this has created discussions internally at Aalborg University. What was the difference between problem-based and problem-oriented? The criticism was that problem-based was much narrower in the approach than problem-oriented. Externally, the researchers from Roskilde took the chance to claim—which is surely right—that they were more loyal to the original ideas compared to Aalborg University (Olsen, 2013).

Internationally, it has created confusion to apply the same acronym for two different curriculum systems (Servant-Miklos et al., 2019). In my own writings, I am not always consistent in spelling out the acronym PBL as problem- or project-based learning, and sometimes I use problem-based project work, sometimes problem-oriented projects, and most times problem- and project-based learning. However, I have always been consistent in pointing to both the problem and the project components.

I may have added confusion to the mix, but every curriculum model is a result of a social construction. The questions are: At which abstraction level we should conceptualize, and what do we want to emphasize—commonalities or differences? Hopefully both, but maybe sameness and commonalities are more important at some point in history than the variations and differences. By saying this, I also indicate that the skepticism and fight for new teaching and learning methods were different during the '90s and the '00s, and there was a need to join forces. During the decade from 2010 the acceptance of PBL increased to become more widespread than societal skepticism.

I agree, however, with the authors who want to create clarity of the different concepts as Ginie Servant indicates, "*the parties in this debate to discuss this situation openly. The risk of not doing so is fomenting further confusion and fragmentation in the practice of PBL until PBL means nothing and everything*" (Servant-Miklos, 2020, p. 13). But maybe the concept of PBL is already outdated or it needs a fundamental 360-degree inspection, especially if a PBL approach has to become a response pedagogy to chaos or collapse.

All student-centered pedagogies might need revisions, as what starts as critical learning approaches becomes mainstream at certain times and could become embedded in an instrumental educational discourse. Routines and daily operations can overshadow any original idea of emancipation. There needs to be a vision and ideas relating to a pedagogy, and it should not only be a question of achieving relevant knowledge and competencies but indeed a matter of educating critical citizens who possess the knowledge, competencies, and attitudes to participate in sustainable development of technology and society. Maybe even the emancipatory ideas should be renewed.

4.3 Development of the Curriculum Model

During the 1980s the first graduates from the university came into academic positions at the university and were using the pedagogical knowledge they'd gained during their studies. In the beginning there was a kind of inbreeding, one of the variables that aided the development and survival of the pedagogical model. These pioneers had experienced, in their own bodies, how it was to work in project teams and how motivating and inspiring it could be working on the projects, especially with real-life problems.

They were also first-generation academics who were not brought up with certain academic values and approaches. A large part of the new academic staff had never experienced anything other than the project organization of learning; they trusted the system and dared to develop it further by experimenting with larger teams and other societal themes for the engineering projects. For example, at Aalborg, one year, the theme of working environment would frame first-year students studying electronics, and the next year it could be a theme within infrastructure.

Descriptions of the Aalborg PBL model can be found in many different publications (Kjærsdam & Enemark, 1994; Kolmos & De Graaff, 2015; Kolmos et al., 2004c). The Aalborg model is based on problem-based project work in which approximately half of the students' time is spent on team projects and the other half on traditional lectures. All project work is done in teams, and the same model is mainly used from the first semester through to the completion of the master's degree. During the normal course of a university degree program, teams typically shrink from 6–7 students in the first year to about 2–3 students in the final semester, see Fig. 4.1.

The curriculum consists of a semester approach, where one semester is 30 ECTS, with 1 ECTS equivalent to approximately 28 hours of student work under the European Credit Transfer System. In the early Aalborg model, the 30 ECTS were as distributed between 15 ECTS project work, 7.5 ECTS project unit courses assessed through the project assessment, and 7.5 ECTS basic science courses assessed separately. The basic science courses were called study courses and were not necessarily related to the project. The project unit courses should support the project, which could be a challenge as the students had the freedom to influence the direction of the

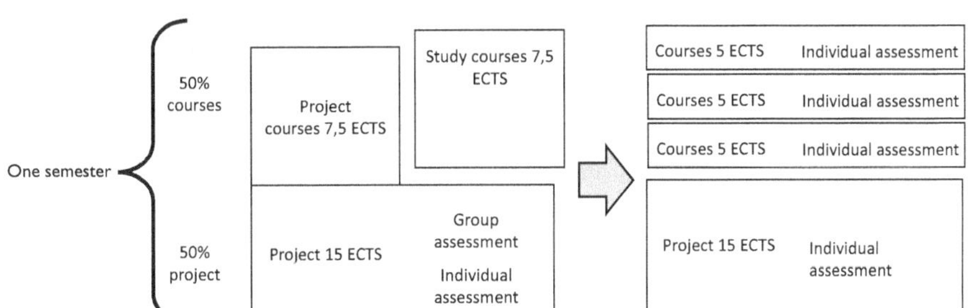

Fig. 4.1 The change of the Aalborg PBL model in 2006

project. If the courses were not relevant for the project, then students were not motivated to participate and learn the content. Approximately three-quarters of the students' time in these courses is dedicated to project work. Only in the latter-mentioned courses are students assessed at the individual level.

At the Faculty for Engineering and Science in 2006 this would change to three single courses of each 5 ECTS and with separate assessment, see Fig. 4.1. The reason for the change was pressure from the subdisciplines in the taught courses as students had a tendency to only pay attention to the content if it could be applied to their projects, and some students discovered they were not accredited in the project exam for the knowledge they had gained in the project courses (Kolmos & Holgaard, 2012).

There is a comprehensive project every semester that provides the possibility to analyze problems and to have the time to relate theory and practice. Deep learning takes time, and research clearly indicates that the learning that students remember after graduation comes from doing projects. In one of the studies, the PROCEED project, we followed all Danish engineering students from their first year to their first job after graduation (Kolmos & Holgaard, 2017; Kolmos et al., 2021; Kolmos & Koretke, 2017). That study is very rich on data, and it can be hard to draw just one important conclusion. However, when asking these young graduates what had contributed to their readiness for their first job, project work comes out as number one, followed by theoretical and lab courses (Kolmos & Koretke, 2017).

Even if all the research results are in favor of the curriculum model and that students indicate that they achieve high competency level, I have always internally criticized the model for being too stiff. Students know how to start up a project, to finish one, to share the work, to keep each other updated, to present at the project exams, and how to liaise with the facilitator, etc. The semester project becomes a routine just like the classroom teaching. What further determines the routine aspects is that the students tend to stick together in the project teams with their complimentary knowledge and skills. Therefore, I have advocated for variation in length and scope of the projects and that students should not be allowed to form the project teams themselves every semester. Variation and comparisons are some of the elements in learning and this can be facilitated by curriculum design.

Clearly, the transition from a traditional teaching model to a PBL model, such as the one used at Aalborg University, is a significant change. The number of contact hours is drastically reduced, and project work not only consumes a significant amount of the students' time, but also serves as the central axis of learning.

4.4 The Methods Approach

Many books were published on more concrete methods and tools to carry out problem analyses and collaborative project processes after the 1970s. At that time the Danish high schools, or K12, were far from having active and collaborative learning methods, and the students needed more guidance in how to approach societal and collaborative learning. Books on theory and methods were just one of the new approaches in the theory of science with a more pragmatic perspective (Adolphsen & Qvist, 1995; Kuada, 2015). There was ongoing criticism on problem-oriented project work for being too pragmatic and that students only learned a selection of the disciplines that were relevant for the analyses and solutions to problems. The students therefore lacked an overview of disciplinary knowledge.

In discussion, the argumentation against this statement was more that the taught courses would provide the overview of the discipline and that the students studying elected parts of the discipline learned at a deeper level when they had learned to apply the knowledge. A third argument was the learning to learn approach embedded in the problem-oriented project work. The criticism still exists and is one of the core obstacles that academic staff have for their trust in PBL. See more about this in Chap. 5.

The problem-oriented project work turned out to be a societal success, not because it fulfilled the embedded intentions of political radicalism, but indeed because the students learned some of the needed skills and competencies, such as collaboration, project management, flexibility, problem analyses, creativity, and innovation. In the wake of the first books on problem-oriented project work, a series of methods books came with the intention of guiding project processes. Already in the early 1980s there were several shorter texts written for students; however, nearly all that material is 'grey' and has not been referenced as it was not fully published (Algreen-Ussing et al., 1978).

As the project methods was applied in many different disciplines, we see project guides to health, veterinary science, engineering, social science, etc. (Algreen-Ussing & Fruensgaard, 1992; Andersen & Schwencke, 2001; Dahl et al., 2005; Mørch, 1993; Thyness, 1987). Quite late in this process of new project work books was a combined tool and method book from Roskilde University with both theory of science, methods and tools (Bitsch Olsen & Pedersen, 2008).

In engineering education at Aalborg University, Helle Algreen-Ussing, Niels Fruensgaard and Bendix Skov-Petersen had taken the task to write an introduction to project work. They met at the first-year program, which at that time was a common year for all special study programs and where the students would select their study. They came from three different departments and in their first booklet, they had a special chapter on engineering methods together with an overview of project phases and some advice for the students on how to handle collaboration (Algreen-Ussing et al., 1978). As the students could choose any program, Helle Algreen-Ussing, Niels Fruensgaard and Bendix Skov-Petersen started to work on the commonalities across the engineering programs. What was unique was the attempt to develop an understanding of engineering as method based, and that it was these methods that could be transferred to other areas.

In social science, there were also booklets written on project work, but these were general terms to understand why this methodology was important; Eva Hultengren reports directly that the writing from Helle Algreen-Ussing and colleagues was much more tool oriented (Hultengren, 1979). Helle Algreen colleagues wrote several books over the years and perhaps the approach began with tools but ended with methods, which have been an important contribution in developing the understanding of what PBL in engineering is. Furthermore, what is more important is that the methods and methodologies can be learned in one project and applied in another—especially problem-solving methods.

4.5 The Problem and the Project

Students need to learn methods for analyzing problems and identifying the specific problems they want to work on and solve (Holgaard et al., 2017). Students must understand the problem in a context, which is the learning objectives in the curricula as well as the specific societal theme they are investigating. Within the theme, there are clusters or areas of problems, and it is among these problems that the students need to choose an area and identify a problem. After the problem-solving phase, the students again need to provide perspectives; see Fig. 4.2 (Holgaard et al., 2017).

Going through these first phases involves certain methods and Holgaard and colleagues have emphasized the '5W1H' model to help students to get a hold on the problem: What are we dealing with? Why is this a relevant and potential problem? Who are the stakeholders involved having different perspectives? Where do we locate the problem? When is it a problem and How can the problem be formulated? The 5W model supports the students in relating the problem to the context.

Just as there are variations in problems, the project also can have many variations. First and foremost, the problem determines the project and the size and length of it. The project is the organization of the learning process, and a project is by definition a unique temporary collaborative endeavor for specific activities, which can be analyzing or solving specific issues and coming up with a conclusion, product or solution (Algreen-Ussing & Fruensgaard, 1992). Many product activities are

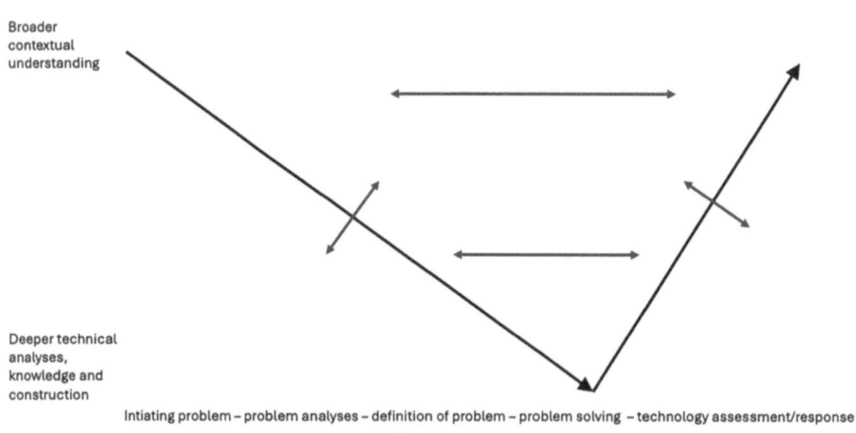

Fig. 4.2 Project phases and relation to types of knowledge

projects from a pragmatic standpoint since they serve a function and have a deadline to complete the product.

The uniqueness and specific activities are related to the problem, which can have many different shapes as already described and represents stakeholders/actors, needs and values. The problem forms the purpose of the project. The size and length of the project depend on the problem that needs to be solved. The size can include a huge variation from narrow discipline projects to more complex mega projects, and the length from shorter to longer courses (weeks, semesters, years) (Kolmos et al., 2020).

In an earlier article, I analyzed the types of problems and projects that were practiced during the 1980s and before and came up with three types of projects (Kolmos, 1996):

- Assignment/task projects with a defined problem, although rarely used.
- Discipline projects with possibility to identify problems within disciplinary boundaries, which are the most often used.
- Problem projects with authentic problems sometimes leading to interdisciplinary projects.

I must confess that I deeply regret the definition of the assignment/task project and have tried to remove that from the project approach many times. At the institutions where I organized workshops, the academic staff there running traditional courses with smaller group work on pre-defined tasks maintained that they were also doing PBL. This is one of the aspects Ginie Servant warned about—that PBL can be everything and nothing. Doing just lab work and having small groups working on assignments is <u>not</u> PBL in my understanding. That is in general more active learning in the classroom and applying active learning strategies can be very efficient for learning (Felder & Brent, 2009; Prince & Felder, 2006).

A project is always temporary and it has a time limit, with a beginning and an ending that involves both project management and leadership, which are explicit, and required competencies for engineers (Routhe et al., 2024). The project organization is the core curriculum element in learning generic competencies such as collaboration, communication, project management and leadership; see Chap. 6. The team size for collaborative projects might vary from smaller groups of 3 or 4 students to larger teams of 8–12.

A team-based organization may consist of one team or a team of teams depending on the problem or purpose of the project, which includes both intra-team and inter-team competencies. These two team perspectives are different, as close collaboration is significant within the team, and the project management and leadership will dominate the collaboration among the teams. However, the inter-team will rely on the intra-team skills and competencies; see more about this in Chap. 7 on interdisciplinary learning.

Each team member brings unique experiences, perspectives, countries, and genders, and the addition of a new member will impact team collaboration. There are different modes for team collaboration patterns. What we normally see are collaboration patterns between real integrated teamwork and more instrumental sharing of work with each individual team member responsible for their own tasks and chapter in the report.

The forming of teams is a curriculum activity and, in principle, there are different ways of forming them (Bundgaard et al., 2021):

- Students organize themselves based on relationships.
- Subjects or topic driven model where the student discusses problems from an interest point of view.
- Administrative teams organized by the academic staff and can be based on learning preferences.

Over the past decade, there has been a demand for the incorporation of complex problems into engineering education through increasingly extensive project frameworks. In Chap. 7, I will present a recently published framework on interdisciplinary projects.

4.6 Space and Facilitation

Aalborg University was established with the philosophy that studying should be similar to working. Each project team had its own room where the students worked and nearly lived for periods when they were finishing their projects. They came in the morning for coursework or project work and used the space as some kind of office space. In engineering education the students came in around 8.00 a.m. and worked until 4 p.m. Having the group room was an identity for the students and a symbol of their collaborative work. Schedules for tasks, times and division of work were on the walls. Many group rooms were like homes where the students brought personal items to decorate the space. The room framed the collective identity for the team and was an important element in the success of the PBL curriculum at Aalborg University.

One room per project team is no longer the trend. Students now have to book shared spaces and they combine various versions of digital platforms (Sørensen, 2022). Open office space is the mantra for new buildings, which in principle should be efficient for networking and knowledge sharing. During the two decades from 2000 to 2015, many universities rebuilt or built space to support a more collaborative learning environment. Having space matters including technological space and systems to support collaborative learning in blended learning forms (Brooks, 2011; Ellis & Goodyear, 2016).

4.7 Assessment

From the beginning, the assessment in the Aalborg PBL model was a team-based assessment or we called it group assessment. Because students have been working in teams to complete their projects since the beginning of their studies, assessing project skills in a group setting is a natural continuation of the PBL-philosophy and corresponds with the theory of alignment (Kolmos et al., 2004c). The project teams submitted a written report and in the assessment session they gave an oral presentation about their project. Following that, the facilitators and external examiners evaluated their communication skills and asked clarifying questions about the oral presentation. From there, the facilitators and external examiners asked the team members questions about their report, and the team members had the opportunity to raise their hand and, if asked, answer the question or elaborate on a point made by one of their peers. Each team member received an individual grade at the end.

The report served as a reference for determining the grades, combined with the individual oral presentation and performance during the questioning session. The facilitator was in charge of the assessment and ensured that the questions asked were relevant to the content of the report or oral presentation.

A total shock came in 2006 when the Ministry of Science, Technology and Innovation and the Ministry of Education issues a press release:

> We have wished to secure that the student has the ability to work in a way which as much as possible correspond to the situation on the labor market. That is, on the one hand to collaborate with others, for example in doing project work and preparing for an exam, and on the other hand by individual performances, where the responsibility is placed on the individual, and that is what characterizes a real exam [Translated]. (Kolmos & Holgaard, 2007, p. 8)

The government was criticized for abolishing group-based exams solely for ideological reasons, citing a lack of scientific evidence to support the decision, which sparked a heated political debate in public media and at higher education institutions.

This created a situation where, as researchers, we could compare two assessment systems for team-projects; however, it also created an issue for the PBL system. If the project teams did not get a collective response to their projects and were not able to show collaborative competencies with peers but only individual grades, we were concerned how this would impact the learning environment. The PBL culture and support at that time were so strong among both academics and students that the change in the assessment system did not influence the opinion of PBL. In 2012, after six years, the group assessment was again legal.

All our studies indicated a split between the academic staff. One third were more or less satisfied with the new restrictions and two thirds were more in favor of the previous system prior to 2006, which was implemented again in 2012 (Holgaard & Kolmos, 2009). Attitudes to the assessment systems are only superficial measurements for the more general attitude to PBL, but it indicated that not the entire institution was aligned in attitudes and opinions.

4.8 Loyalty and Concerns from the Academic Staff

When you have helped to build something, your loyalty will be more apparent, and especially in new universities or organizations you feel like a pioneer. That also counts for the academic staff at Aalborg University. Their loyalty has been exceptional which is often forgotten. Perhaps that loyalty was also based on the fact that many of the academic staff employed during the 1980s had been educated at Aalborg University and were first generation academics. As universities expanded during the 1970s and '80s, there would be more students without academic familiarity. The narrative was that the PBL system actually helped get first generation students into academia. There is no proof of this, however. We only have the fact that during the first many years, Aalborg University recruited these students primarily from a region that earlier had been without a university. But for these first- generation students, PBL was the academic norm and they trusted the system.

International interest increased for Aalborg University and this made the academic staff proud. During the 1990s we had visitors from Stanford, where Professor Larry Leifer said with surprise: "*Here they have made an exceptional system, and I am not sure that the faculty are aware of the uniqueness they have created.*" In this respect, the UNESCO Chair and later the UNESCO PBL Centre with all the visitors was an external acknowledgment and recognition of the PBL system. The academic staff were particularly proud when Ruth Graham published the MIT report in which Aalborg University was ranked as the internationally fourth most inspirational university for innovations in engineering education (Graham, 2018).

The book *The Aalborg PBL Model—Progress, Diversity and Challenges* published in 2004 has contributions from all faculties at the time: humanities, social science, engineering, and science (Kolmos et al., 2004a, 2004b). The last article in the book was written from a professor originally from Germany who was honest enough to admit his difficulties in adjusting to a PBL model based on values from the 1970s. He emphasized the distances in power structures, which have always been larger in the German society compared to the Danish. Many of the newcomers with an educational background from other universities had been mistrustful in the beginning but were convinced by the practices at the departments.

Visions and concerns were collected among the academic staff in a working group formed by the vice rector in 2015. Being responsible for the engineering part, I asked for challenges, wishes for 2020, and specific activities; these responses were analyzed but never published. Seventeen responses from the engineering staff witnessed a strong commitment. The results from these responses were not aligned in the same direction; however there were some common points:

The first concern was that resources for teaching and learning have decreased. Aalborg University receives the same public funding for each teaching credit as all the other universities, but what might differ is the internal distribution of resources at different institutions. Resources also have to do with space—so this was a place to go from each project team having their own project room to shared rooms or no project rooms at all. Furthermore, resources include time for facilitation of the project teams. From the beginning of 2015, there has been a decrease in hours for a project team of 6–8 students from nearly 140 hours to 80 hours.

The facilitator was another important factor. If the facilitator does not have the time to talk to the students, to have dialogues on issues in the project design and methodology or in the accurate calculations, there is a missing link to academic motivation. Academic motivation may increase when you see and hear others' enthusiasm and logical reasoning. You learn from the dialogue with the academic facilitator who uses the time to discuss and ask the difficult questions, and in an open-minded way identify new ways together with the students. This is motivating for both students and academic staff.

The second concern came after the curriculum reform in 2006, where another course structure was implemented and a disconnect between projects and course units then became an issue. As with the group assessment, the relationship between taught courses and projects in the beginning were continued because of cultural habits and academic staff who emphasized the learning in the projects. One of the concerns in 2015 was that the disconnect starting between projects and courses. Just as PBL survived the change in the assessment system, the system and the culture protected the projects in the beginning by keeping the relationship steady between taught study courses and projects. However, after 10 years in 2015 this was no longer the case. The assessment situation had been solved, but the possible disconnection between courses and projects was not. Therefore, a concern from this study indicated that the courses and projects should be reconnected.

The third concern was that some disciplines were also moving away from PBL principles. For example, the funding to support project spaces and facilitation in the humanities and social sciences was limited. The application of PBL should be strengthened in all disciplines; it was then suggested that the PBL competencies should be more visible in all study regulations. New activities were initiated following this evaluation period, e.g. PBL principles embedded in the internal accreditation system, the establishment of a more explicit learning of PBL competencies, and a research-based PBL vision including digitalization—the PBL future (Kolmos & Ryberg, 2023).

4.9 The Paradox of Institutional PBL Principles

Around 2009, the rector of Aalborg University formed a team of international and internal specialists to develop the basic AAU PBL concepts. International specialists interviewed PBL researchers, staff, and students at the university to acquire a comprehensive understanding of PBL practices with the objective of reformulating the curriculum principles (Barge, 2010). At the time, the objective was to more explicitly define the overarching governing PBL principles spanning science, engineering, health, humanities, and social science, which could serve as internal and external recommendations for PBL curriculum (Kolmos, 2013). The AAU PBL principles have evolved over time, and these are currently under revision (in 2025).

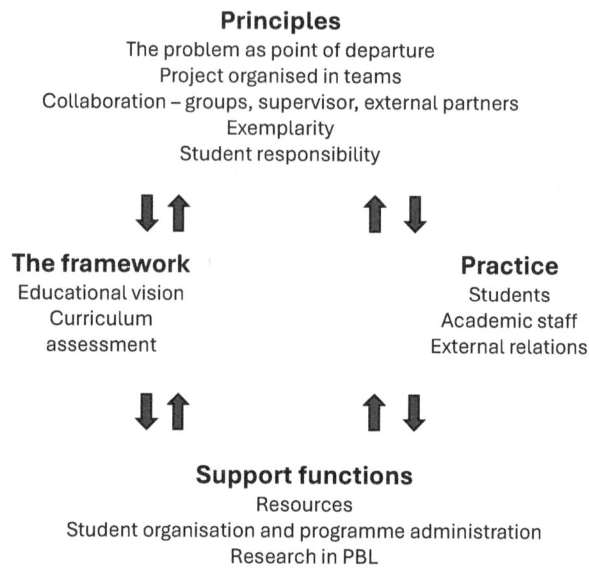

Fig. 4.3 Aalborg University PBL principles

For the 2015 version, there are four headings: principles, framework, practice, and support functions; see Fig. 4.3 (Aalborg University, 2015).

The fact that an institution formulates pedagogical principles is a strong pedagogical statement. There is no doubt that the international recognition that the engineering education received was an internal enabler of a stronger hold in the development. From my experience, engineering has been the most loyal PBL practitioner, whereas humanities and social sciences slowly went in a more traditional direction. The intention to formulate principles for all faculties was to establish a common model; however, if the resources are not aligned with the intentions, it might stay at an intentional level.

Internally, these principles gradually became the guiding principles for all curricula at Aalborg University, particularly after the university had implemented an auditing system that required it to develop institutional criteria for curriculum development (Habbal et al., 2024). In that regard, the degree of formal institutionalization of PBL has expanded since the university's inception in 1974. In addition to incorporating PBL concepts into the accreditation system, the University further institutionalized them by producing an educational profile defining problem-based project work, research-based education, collaboration, and student-centered learning as fundamental values.

The University also established educational program targets based on the core attributes required of university graduates, stating that educational programs must foster graduates who can perform problem-solving work, have a solid scientific knowledge foundation, collaborate, and are able to work in interdisciplinary contexts. Because of the link between PBL and institutional accreditation, all programs have to demonstrate that the PBL principles are met and reflect the educational profile.

When learning principles becomes part of accreditation systems, there are advantages and disadvantages. During my career as an international change agent for engineering education, I argued to integrate new experiments into the formal curriculum, as the students would get formal credits. But to do this and then use it as a kind of control mechanism for how PBL-minded an education is could be against the intention of PBL itself. Therefore, the disadvantage is that what should have been an emancipatory pedagogy is applied as a means of "oppression". If it is used as criteria for benchmarking it would be great, but if it would be used as a 'bureaucratic function' in a system, this does not align with the intention of PBL. It definitely takes the romance away, which is the ideology of emancipation of the self and the collective by letting the students follow their experiences and interests within some given frames, and a romance for academic staff in the freedom to experiment and use imagination of alternatives and what could be. Somehow, to protect the PBL system, even with good intentions we might end up in bureaucratic regulation of the curriculum. Sometimes academic staff feel this as restrictions, and this is a paradox and a fundamental misalignment.

As an explanation, a major part of the regulation originates from the political level. However, institutionalization often hinders development, as e.g., accreditation criteria in many countries can be barriers for moving to more ill-structured project work as there would not be a fixed syllabus.

Aalborg University has changed a great deal and embedded in this change are institutional structures, cultures, priorities, and pedagogical attitudes. This change also comes together with a priority on research, which has been a common trend globally and where Aalborg University also has performed rather well internationally.

The priority of research comes with a prize and the romance that has existed around the Aalborg PBL model might have transformed for the actors involved. Romance belongs to happiness and intuition with space for experiments and a feeling that everything is possible. In contrast to this is realism and a more instrumental approach, which occur automatically during periods of an increase in student numbers, simultaneously with a funding decrease. Generally, there has been a decrease in financial resources for university teaching, and this has been nearly constant over the last 30 years, which creates constant pressure.

4.10 Variation in the Problem-Based Project Curriculum

There will be variations no matter which PBL practice we analyze. In the case-based PBL model, there are open and narrow cases and similar ones for the project-based PBL. Erik de Graaff and I started to work on a model with a variation in the PBL curriculum (see Table 4.1). Our message was clear to participants in workshops and the popular visitor workshops that we established at the UNESCO Chair

Table 4.1 Dimensions in a PBL curriculum based on the Aalborg PBL model (Kolmos et al., 2009)

Curriculum element	Discipline and teacher-controlled approach	Innovative and learner-centered approach
Objectives and knowledge	Traditional discipline objectives	PBL and methodological objectives
	Disciplinary knowledge	Interdisciplinary knowledge
Type of problems and projects	Narrow	Open
	Well-defined problems	Ill-defined problems
	Discipline projects	Problem projects
	Study projects	Innovation projects
	Lectures determine the project	Lectures support the project
Progression, size and duration	No visible progression	Visible and clear progression
	Minor part of the curriculum	Major part of course/curriculum
Students' learning	No supporting courses	Supporting courses
	Acquisition of knowledge	Construction of knowledge
	Collaboration for individual learning	Collaboration for innovation
Academic staff and facilitation	No training	Training courses
	Teacher-controlled supervision	Facilitator/process guide
Space and organization	Administration from traditional course- and lecture-based curriculum	Administration supports PBL curriculum
	Traditional library structure	Library to support PBL
	Lecture rooms	Physical space to facilitate teamwork
Assessment of evaluation	Individual assessment	Group assessment
	Summative assessment	Formative assessment

and later continued under the umbrella of the UNESCO PBL Centre. You can gain inspiration by learning about the Aalborg PBL model, but you can never 'copy paste'. There are so many variables of which some are tacit and cultural in the function of a curriculum, but there are also core elements that are necessary to address when moving to more active learning and, in particular, PBL.

In a curriculum, there needs to be alignment among the elements. In Table 4.1 some of the curriculum elements are mentioned with extremes of two approaches: teacher-controlled and innovative/learner-centered. Many points exist between each extreme dimension, and any curriculum piece may include multiple dimensions. The purpose of this is to illustrate the possibilities in a curriculum construction—partly by outlining the variables included in curriculum considerations and partly by pointing to possibilities. There might be many more possibilities than mentioned, and it is not the point to cover the whole but to open the mind for choices.

PBL methods vary greatly and there are many possibilities, for example: the length of projects as one semester or one week; the size of the team; how to handle group formation processes; the interaction among all the curriculum elements; the interaction with society in identifying problems; the culture and the national politics for higher education, etc. (Kolmos, 2010).

4.11 Critical Voices

Numerous Danish and Nordic critiques emerged over the problem-oriented project work that developed throughout the 1970s. In Danish higher education research, project pedagogy was the primary focus of discussion and documentation. During that period, the novel and pivotal educational approach was examined by Karen Borgnakke, who analyzed the contemporaneous discussions and discourses in the social sciences (Borgnakke, 1983). One was either in favor of or opposed to, with no intermediate viewpoints.

Karen Borgnakke indicates that the most severe criticism originated from certain Nordic colleagues, specifically Erling Lars Dale, professor in Norway, and Steinar Kvale, former professor at Århus University (Borgnakke, 2011). They criticized the project pedagogy for contributing to social anomie, fostering narcissistic personality development, and resulting in professional stagnation and disqualification. A distinctly incisive voice that is rooted in the emphasis on participative principles and student-centeredness in learning would, in their view, not lead to professionalism.

The same critique was formulated internally at Aalborg University as a reaction to the students' ownership of the projects and especially the formulation of the problem. The message from one of the professors at Social Science, Per Salomonsen, was that experience leads to ignorance of scientific knowledge, and the students do not learn any scientific or professional knowledge. Students would remain at their experiential knowledge level without any progression or learning.

Karen Borgnakke, who experienced the battlefield at seminars and conferences, concludes that the critique was beneficial because it strengthened the scientific argument for the project pedagogy at that time. In a Danish doctoral thesis degree, she reconstructed the argumentation and an understanding of the scientific and professional approach in the problem-oriented work (Borgnakke, 1983).

Following the ideological and theoretical struggles of the 1980s, empirical research was beginning to get published. One of the studies was on students' experiences with project facilitators at Roskilde University, where it openly stated that some of the student project teams did not use their facilitator since they felt he or she could not contribute (Christensen, 2008). In engineering education, I am sure you will be able to find student project teams that are not really using their facilitator; however, the scientific knowledge is quite different from the social sciences, and the experiential part can merely be identified in the problem analyses phases which is contextual, whereas the solution phase applies technical scientific knowledge, where facilitator input can be very valuable. Therefore, the battles were never the same in engineering education as in social science and humanities.

As a former social science student, I would add to some of the criticism that project-based learning can become overly pragmatic and elective. Many students work with the same team members from project to project, honing their skills in data collection, theory formulation, coding, and other areas. Furthermore, students tend to apply what they already know and if there is no drive to learn new ideas or methodologies, they may miss out on the wealth of theories and learning opportunities available. All of this can be addressed by incorporating learning outcomes and providing overviews throughout the taught courses. However, there is a hidden curriculum in PBL and all other curricula, which must be discussed openly. The moment for defense has long passed.

The same critique was formulated of the case-based PBL in another battlefield, but with more or less the same argument. A very often cited critique of PBL is the one by Kirschner, Sweller, and Clark. Their point is that minimal guided instructions are less efficient than teaching approaches with rich guidance, and there is research documenting that guidance from experts increase students' learning (Kirschner et al., 2006). This article was published in 2006, and the level of guidance can hardly be the only variable in students' learning efficiency. I fully agree that resources for teaching and learning are important; however, the decrease in resources for the last 20 years at universities for teaching and research, with a consequence for the resources for counseling, supervision, facilitation, or guided instruction, would mean that student learning efficiency today should have also decreased noticeably.

At Aalborg University, there is an increasing internal critique of the existing system. Thomas Ryberg recently raised a concern that collaborative learning and team project processes in particular are aiming for consensus among the team members. He argues that there is a value in working with difference and consent instead of consensus, as consent is less inevitable for the individual student (Ryberg, 2024). Within engineering education, consensus is a must when the students are constructing devices together.

The flexibility in curriculum and students' competency profiles are addressed by one of my colleagues from the social sciences (Turcan & Reilly, 2020). The authors analyze the evolution of graduates' competency profiles from 2015 to 2030, where they see a move from complex problem solving, coordination, people management, negotiation, and active listening to a new list of competencies in 2030 that include mental elasticity and complex problem solving, interdisciplinary knowledge, critical thinking, creativity, and people skills among some of the core skills. Turcan (2020) concludes in his chapter:

> We live in liquid times, characterized by dynamic change, high degree of newness, innovative power, and negative social impact, a politics of managed ignorance, fear, and science denial. The liquid world affects free will, moral values, and reconfigures skills and competences. Graduates need to be equipped to navigate this world and to exercise an informed and critical judgement and evaluation. One way in which they can be prepared through PBL is to: become context-free; think and step outside the box, see a big picture and reframe the problem or the question; learn and apply new methodological and/or theoretical lenses; employ new, untried research methods and techniques; rearrange things to find a solution and create new knowledge. This can be achieved by developing a strong interdisciplinary knowledge base; mental elasticity; complex problem solving; critical thinking; creativity; people skills, STEM and SMAC[1] skills. (Turcan, 2020, p. 417)

[1] STEM is Science, Technology, Engineering and Mathematics. SMAC is *social, mobile, analytics, and cloud.*

The STEM and SMAC skills should be understood not as isolated skills on the list but more integrated with elasticity, complexity, etc. In a research project funded by internal strategic funds from Aalborg University—the PBL-Future project—there is an embedded critique of the PBL as it has played out during the years. For the future direction, this research project calls for variation in the curriculum, flexibility in the creation of students' learning tracks, and integration of digitalization in a hybrid mode (Kolmos & Ryberg, 2023). In the Aalborg system, it is hard to create individual learning trajectories as the three courses are determined, and even if there is freedom in the scope of the project, the choices will always be influenced and determined by the project team.

Almost at the same time, another book was published with Aalborg University called *The Bildung potential (educational) from project work* [translated title] (Feilberg et al., 2022). Casper Feilberg is a PBL researcher with a background in psychology and his perspective is not to merge but to combine the lenses of educational theory (*Bildung*), pedagogical theory of project work and narrative theory. The learning process is seen as a personal-disciplinary learning process in which the individual builds up not only the disciplinary knowledge but the internalization of *Bildung*, a societal value. He argues that ethical-political education and personal-professional education are important aspects of students' learning processes in higher education. These aspects are under pressure as the formal educational system with explicit learning outcomes dominate the informal system. Feilberg tries to update the values of the original PBL principles in pointing to the need for and potential of learning values. His context is humanities and social science.

The same context is the frame for another young scholar from Roskilde University who made a discourse analysis of some origins and core documents of the establishment of the Danish project pedagogy (Sørensen, 2023). He argues that the discourses have changed dramatically from the emancipatory learning approaches to new knowledge economy with competency rationale. He maintains that the Humboldtian view of the role and function of a university should be revitalized in a new development based on pedagogical and philosophical perspectives in future reconstructions.

One of my former PhD students recommends reading Illeris again and he is surprised that this iconic reference for problem-oriented project work or the Danish inspiration to PBL, which embrace both problems and projects, is both practical and contemporary in pointing at lifelong learning, employability and learning to learn as core principles of the origins (Boelt, 2024).

Thus, there are critical voices of the PBL system historically as well as contemporarily. However, none of these voices use the wording of the challenge of chaos and young people's anxiety of the future. The criticism does not include any mention of the lack of emotional aspects in learning but highlights the lack of emancipatory and empowerment aspects together with the lack of flexibility, which are necessary components in future PBL.

4.12 Reflection 4: Living-With Chaos

There is no doubt that Aalborg University has had a tremendous impact on the change process of engineering education in particular; however, the PBL model itself has also changed from being rooted in a more critical discourse to a much more instrumental learning approach. After 50 years, the pioneers who created the original culture are no longer around and new generations of academics have taken over with formulating specific learning outcomes in the curriculum and with a priority for research. These have been the same conditions for all universities.

Surely, culture and curriculum must change during a span of 50 years, with many system critiques along the way together with a change in political systems. However, what has been part of the pioneering culture is the general enthusiasm, engagement, involvement, and commitment. It has been a collective effort to build up a new type of pedagogy; even if it has been during different conditions, it witnesses a courage to create alternatives. Maybe the enthusiasm and engagement are more limited today as there have been changes every year and there is a group of academic staff who are exhausted.

PBL has an advantage in responding to chaos as students learn to analyze problems and identify problems. This can be a potential for integrating chaotic problems into the curriculum. However, we also have to be aware that if intended emancipatory learning becomes part of a control system, the pedagogy loses its' potential. Enthusiasm and involvement among staff will be needed for curriculum development.

From the cognitive side, students need to analyze problems from different angles and assess possible implications. This means that in this technology-society, social justice and ethical angles need to be addressed as part of the curriculum. Risk management should be highlighted to a much higher degree as there will be risks with every technological development and the students should be trained in that way of thinking.

The cognitive part is not enough, however. We need to start talking about anxiety and chaos and the ability to immediate reactions. I am not sure this is what the younger generation of PBL researchers are doing; however, some have started as they draw on the emancipatory part of learning and want to make this alive again. Emancipation in the 2020s might be very different from emancipation in the 1960s and 1970s and I think one of the tasks is to revitalize this again with a new agenda and new perspectives.

Anxiety is something we all feel, not only the younger generation, and we need to see hope and be able to master things instead of giving up. To master your own learning, you need to have the ability to make decisions, to identify emerging problems, share ideas, and learn to trust other contributions. PBL needs to be reformulated for a new agenda and in a new context with both collaborative cognitive and emotional strategies.

References

Aalborg University. (2015). *PBL problem-based learning*. http://www.aau.dk/digitalAssets/148/148025_pbl-aalborg-model_uk.pdf

Adolphsen, J. B., & Qvist, P. (1995). *ABC i problemformulering, problemløsning og projektskrivning*. Gyldendal.

Algreen-Ussing, H., & Fruensgaard, N. O. (1992). *Metode i projektarbejdet: problemorientering og gruppearbejde* (3. opl. ed.). Aalborg Universitetsforlag.

Algreen-Ussing, H., Fruensgaard, N. O., & Skov-Petersen, B. (1978). Projekt og rapport: rapportskrivning i projektorganiseret undervisning. *Aalborg*. Universitetsforlag.

Andersen, E. S., & Schwencke, E. (2001). *Projektarbejde - en vejledning for studerende*. Studenterlitteratur.

Barge, S. (2010). *Principles of problem and project based learning: The Aalborg PBL model*. Aalborg University.

Bitsch Olsen, P., & Pedersen, K. (2008). *Problemorienteret projektarbejde: en værktøjsbog* (3. udgave, 2003, 4. oplag. ed.). Roskilde Universitetsforlag.

Boelt, A. M. (2024). A personal introduction to Knud Illeris' Problemorientering og Deltagerstyring–Oplæg til en Alternativ Didaktik. *Journal of Problem Based Learning in Higher Education, 12*(2), 34–41.

Borgnakke, K. (1983). Projektpædagogikken gennem teori og praksis. *Aalborg*. Universitetsforlag.

Borgnakke, K. (2011). *Et universitet er et sted, der forsker i alt–undtagen i sig selv og sin egen virksomhed. Rapport om den forskningsfaglige baggrund for udvikling af universitetspædagogisk forskning*. Det Humanistiske Fakultet. Københavns Universitet.

Brooks, D. C. (2011). Space matters: The impact of formal learning environments on student learning. *British Journal of Educational Technology, 42*(5), 719–726.

Bundgaard, S. B., Dahl, B., Krogh, L., & Ravn, O. (2021). Exploring PBL group formation processes. *Aalborg*. Universitetsforlag.

Christensen, G. (2008). Vejledning–mellem magt og afmagt. *Dansk Universitetspædagogisk Tidsskrift, 3*(5), 4–9.

Dahl, A., Dich, T., Hansen, T., & Olsen, V. (2005). *Styrk projektarbejdet - en redskabsbog til problemorienteret projektarbejde*. Biofolia.

De Graaff, E., & Kolmos, A. (2003). Characteristics of problem-based learning. *International Journal of Engineering Education, 19*(5), 657–662.

Ellis, R. A., & Goodyear, P. (2016). Models of learning space: Integrating research on space, place and learning in higher education. *Review of Education, 4*(2), 149–191.

Feilberg, C., Bo, I. G., & Romani, A. Q. (2022). *Projektarbejdets dannelsespotentiale*. Aalborg Universitetsforlag.

Felder, R. M., & Brent, R. (2009). Active learning: An introduction. *ASQ higher education brief, 2*(4), 1–5.

Graham, R. (2018). *The global state of the art in engineering education*. M. New Engineering Education Transformation. neet.mit.edu.

Habbal, F., Kolmos, A., Hadgraft, R. G., Holgaard, J. E., & Reda, K. (2024). *Reshaping engineering education: Addressing complex human challenges*. Springer Nature.

Holgaard, J. E., & Kolmos, A. (2009). Group or individual assessment in engineering, science and health education. In A. Kolmos & A. Kolmos (Eds.), *Research on PBL practice in engineering education* (pp. 57–69). Sense Publishers.

Holgaard, J. E., Guerra, A., & Kolmos, A. (2017). Getting a hold on the problem in a problem-based learning environment. *International Journal of Engineering Education, 33*(3), 1070–1085.

Hultengren, E. (1979). *Tværfaglig undervisning som politisk undervisning*. Institut for Uddannelse og Socialisering.

Jensen, S. A., & Wagner, M. F. (1990). *Ingeniørers arbejde og kvalifikationer - en foreløbig undersøgelse af arbejds- og kvalifikationsforhold blandt AUC-ingeniører*. A. Institut for Elektroniske Systemer.

Kirschner, P. A., Sweller, J., & Clark, R. E. (2006). Why minimal guidance during instruction does not work: An analysis of the failure of constructivist, discovery, problem-based, experiential, and inquiry-based teaching. *Educational Psychologist, 42*(2), 115–121.

Kjærsdam, F., & Enemark, S. (1994). *The Aalborg experiment project innovation in university education*. Aalborg Universitetsforlag.

Kolmos, A. (1996). Reflections on project work and problem-based learning. *European Journal of Engineering Education, 21*(2), 141–148. https://doi.org/10.1080/03043799608923397

Kolmos, A. (2010). Premises for changing to PBL. *International Journal for the Scholarship of Teaching & Learning, 4*, 1–7.

Kolmos, A. (2013). Problem- and project-based learning in a global perspective. In A. Aarup Jensen (Ed.), *Visions, challenges and strategies* (pp. 47–66). Aalborg Universitetsforlag.

Kolmos, A., & De Graaff, E. (2015). Problem-based and project-based learning in engineering education: Merging models. In *Cambridge handbook of engineering education research* (pp. 141–160). Cambridge University Press. https://doi.org/10.1017/CBO9781139013451.012

Kolmos, A., & Holgaard, J. E. (2007). *Delrapport 1: Sammenligning af gruppebaseret og individuel eksamen sommer 2006 ved Fakultet for Ingeniør-, Natur- og Sundhedsvidenskab* [Rapport].

Kolmos, A., & Holgaard, J. E. (2012). *Evaluering af ændringerne i PBL modellen på TEKNAT AAU*.

Kolmos, A., & Holgaard, J. E. (2017). Impact of PBL and company interaction on the transition from engineering education to work. In PBL, Social Progress and Sustainability, *Aalborg University Press*.

Kolmos, A., & Koretke, R. B. (2017). *PROCEED-2-WORK Nyuddannede ingeniørers erfaring med overgang fra uddannelse til arbejdsliv: Arbejdsrapport nr. 3*.

Kolmos, A., & Ryberg, T. (2023). *PBL in a digital age*. Aalborg Universitetsforlag.

Kolmos, A., Fink, F., & Krogh, L. (2004a). The Aalborg model. In A. Kolmos, F. Flemming, & K. Lone (Eds.), *The Aalborg model: Progress, diversity and challenges* (pp. 9–18). Aalborg Universitetsforlag.

Kolmos, A., Fink, F., & Krogh, L. (2004b). *The Aalborg PBL model: Progress, diversity & challenges*. Aalborg University Press.

Kolmos, A., Krogh, L., & Fink, F. K. (2004c). *The Aalborg PBL model: Progress, diversity and challenges* (3. printing. ed.). Aalborg University Press.

Kolmos, A., de Graaff, E., & Du, X. (2009). *Diversity of PBL– PBL Learning Principles and Models*. In Research on PBL practice in engineering education. Sense.

Kolmos, A., Bertel, L. B., Holgaard, J. E., & Routhe, H. W. (2020). Project types and complex problem-solving competencies: Towards a conceptual framework. In *Educate for the future: PBL, sustainability and digitalisation 2020* (pp. 56–65). Aalborg Universitetsforlag.

Kolmos, A., Holgaard, J. E., & Clausen, N. R. (2021). Progression of student self-assessed learning outcomes in systemic PBL. *European Journal of Engineering Education, 46*(1), 67–89.

Kuada, J. E. (2015). *Projektskrivningsguiden: Forskningsmetode og projektarbejde*. Samfundslitteratur.

Mørch, S. (1993). *Projektbogen: teori og metode i projektplanlægning*. Rubikon.

Olsen, P. B. (2013). The difference between problem-oriented education and problem based learning: Charactersitics of the regimes of Roskilde University and Aalborg University. In *Visions, challenges and strategies: PBL principles and methodologies in Danish and global perspective* (pp. 335–349). Aalborg Universitetsforlag.

Prince, M. J., & Felder, R. M. (2006). Inductive teaching and learning methods: Definitions, comparisons, and research bases. *Journal of Engineering Education-Washington, 95*(2), 123.

Routhe, H. W., Holgaard, J. E., & Kolmos, A. (2024). Students' learning of management and leadership in engineering education – A literature review. *European Journal of Engineering Education, 49*(3), 540–576.

Ryberg, T. (2024). Dark sides, Tyranny, and Dragons: Attending to the shadows of PBL. *Journal of Problem Based Learning in Higher Education, 12*(2), 89–97.

Servant-Miklos, V. (2020). Problem-oriented project work and problem-based learning: "Mind the gap!". *Interdisciplinary Journal of Problem-Based Learning, 14*(1).
Servant-Miklos, V. F., Norman, G. R., & Schmidt, H. G. (2019). A short intellectual history of problem-based learning. In *The Wiley handbook of problem-based learning* (pp. 3–24). Wiley.
Sørensen, M. T. (2022). *Students' orchestration of groupwork and the role of technology.*
Sørensen, K. A. (2023). *What is the educational purpose of problem-oriented project-based learning?: Reading the discursive construction of the educational aims and purposes of problem-oriented project-based learning in textual introductions from 1974–2018.* Roskilde Universitet.
Thyness, P. (1987). *Prosjekt Undervisning.* TANO.
Turcan, R. V. (2020). Liquid times – Newness and uncertainty: An innovative AAU PBL response. In *Populism and higher education curriculum development: Problem based learning as a mitigating response.* Springer.
Turcan, R. V., & Reilly, J. E. (2020). *Populism and higher education curriculum development: Problem based learning as a mitigating response.* Springer.

Chapter 5
PBL Learning Principles and Learning to Master Collaborative Learning

5.1 Towards an Understanding of PBL Learning Principles

Earlier in the book, I asserted that PBL cannot be understood as a specific model as the disciplines and the institutions have great differences in the conditions for applying more active and student-centered learning. To understand the PBL implemented globally in many disciplines, departments, institutions, and individual courses, we need to formulate this at a more abstract level. The definition of PBL can no longer be at the concrete curriculum level but must be at that the level of learning principles (de Graaff & Kolmos, 2003, 2007). This was the argument to create learning principles as some overall guiding principles covering the many variations of PBL practice.

As indicated in Chap. 3, the concept of PBL has also become less clear, although concrete models should always include variations, particularly when using PBL in various educational systems that represent a diverse spectrum of practices. We created an overview of the PBL learning principles based on foremost practice analyses and on a perspective of learning theories. In the article I wrote together with Erik de Graaff more than 20 years ago, we tried to summarize our practice experiences in three categories: cognitive learning, social learning, and the scientific content, see Fig. 5.1 (de Graaff & Kolmos, 2003).

Cognitive learning addresses the problem as a starting point for the curiosity and engagement in learning. The problem gives a purpose and direction for the learning process and is an anchor that the learning can be related to along the way. There is an embedded theory-practice relationship as well as a theory-experience one that facilitates the understanding of the knowledge and the context in which the knowledge will play out. The project is the frame for the learned knowledge and competencies, as the problem, resources, and time will set the boundaries for the start, process, and end product.

A. Kolmos, *Engineering Education for Interdisciplinary Futures*, Emergent Directions in Engineering Education 2,
https://doi.org/10.1007/978-3-032-32540-2_5

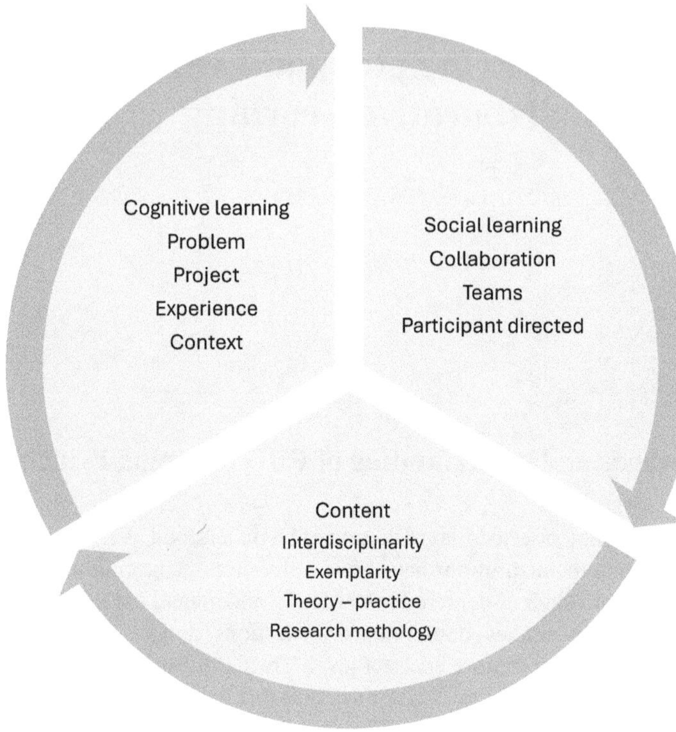

Fig. 5.1 PBL learning principles (de Graaff & Kolmos, 2003)

Social learning focuses on collaboration, team-based learning, and ownership. The team learning component supports the learning process as a social act in which learning occurs through dialogue and communication. Furthermore, students learn from each other, share knowledge, and manage collaborative learning processes. The social approach also includes the concept of participant-directed learning, which emphasizes collective ownership of the learning process and, in particular, problem identification.

The **Content** includes learning outcomes and academic knowledge and involves disciplinary and interdisciplinary learning in particular, bridging traditional disciplinary boundaries and methodologies. It is an exemplary practice in the sense that the learning outcomes are indicative of the general aims, and it strengthens the relationship between theory and practice by including an analytical and methodological approach to problem analysis and problem solving.

In the USA, Cindy Hmelo-Silver was one of the academics formulating case-oriented PBL. She argued that problem-based curricula provide student learning by solving complex, real-world problems by learning from PBL and she formulates the following principles (Hmelo-Silver, 2004):

- construct an extensive and flexible knowledge base
- develop effective problem-solving skills
- develop self-directed lifelong learning skills
- become effective collaborators
- become intrinsically motivated to learn.

In a recent study on the application of PBL in STEM, the intention was to identify core principles for PBL (Smith et al., 2022). Experts in the field, who were researchers or worked at some of the world's leading PBL institutions, were interviewed. The data is rich and has some core elements that we should not forget to emphasize more, e.g. that learning is an active process and a looped process. The facilitation of student project teams must never be linear but also follow the looped learning processes, e.g. how the problem is rephrased many times during a project process. Based on the data analyses, four principles of PBL were defined (Smith et al., 2022):

- flexible knowledge, skills, and capabilities
- active and strategic metacognitive reasoning
- collaboration based on intrinsic motivation
- problems embedded in real and rich contexts.

In this approach, PBL learning principles are formulated at a more general level and, as such, it is not possible to apply the principles directly at a curriculum level.

In trying to create an overview of the principles from the origins to the latest contributions, there are nuances in the vocabulary of the learning strategy, the collaboration, and outcomes; however, the similarities are significantly recognizable. There seems to be an agreement, although various aspects are emphasized and given different concepts, see Table 5.1.

Table 5.1 Comparison of PBL learning principles

	Hmelo-Silver (2004)	de Graaff and Kolmos (2003)	Smith et al. (2022)
Learning strategy	Develop self-directed lifelong learning skills.	Cognitive learning and social learning. Problem, project, experience and context.	Active and strategic metacognitive reasoning. Problems embedded in real and rich contexts.
Knowledge and outcomes	Construct an extensive and flexible knowledge base. Develop effective problem-solving skills.	Content. Interdisciplinarity, exemplarity. Theory and practice. Research methodology.	Flexible knowledge, skills and capabilities.
Motivation and collaboration	Effective collaborators. Intrinsically motivated to learn.	Social learning. Collaboration, teams and participant directed learning.	Collaboration based on intrinsic motivation.

The learning strategy embraces self-directed cognitive learning and active and metacognitive learning, with an emphasis on problems. For the knowledge and outcomes, there are differences in the wording and only de Graaff and I mention here the principles of interdisciplinarity, exemplarity, relationship to practice, and research methodology as something unique. I will address interdisciplinarity in Chaps. 7 and 8, as some of the reform universities were born with an intention of interdisciplinary learning; however, the interdisciplinary focus was less dominant in engineering education. For the motivation and collaboration, the collaborative learning is core to all authors.

What is remarkable, however, is that none of the authors note principles for the involved academics. Barrows did that in his original principles, indicating that there is a new role for the teacher as a facilitator or tutor (Barrows, 1996).

5.2 Similar Student-Centered Learning Principles

5.2.1 Conceive, Design, Implement and Operate: The CDIO Initiative

For engineering education, the CDIO initiative has had similar impact if not more on the change process. The CDIO initiative for engineering education reform started as a project in 2000 by MIT, Chalmers University of Technology, KTH Royal Institute of Technology, and Linköping University (Crawley et al., 2014a; Edström, 2017). Apart from Linköping University, which was founded during the 1970s, all these institutions have a long historic tradition for teaching and learning, with an orientation towards disciplinary courses. The overall goals of the CDIO alliance were formulated by the four founding universities as a pursuit to educate students to master a deep working knowledge of technical fundamentals, to lead in the creation and operation of new products and systems, and to understand the importance of strategic value of their future research work (Bankel et al., 2003; Berggren et al., 2003; Crawley et al., 2014a).

The CDIO initiative has developed 12 standards and a syllabus to guide change, which is formulated within engineering as the professional frame. There are basically four groups of standards: the first group describe criteria for a curriculum with context, integrated design, introduction courses to engineering practice, design and implementation processes, and integrated program evaluation. The second group calls for integrated learning experience, personal and interpersonal skills, and product, process, and system-building skills. The third group concerns workspace and active learning, and the last group is about faculty training (Crawley et al., 2010, 2014b). As such, the CDIO initiative demonstrates a systems approach to curriculum design.

The CDIO syllabus is created from an empirical base combined with inspiration from existing standards. The CDIO syllabus structures the intended knowledge, skills, and attitudes into four themes: (1) technical knowledge and reasoning; (2) personal and professional skills and attributes; (3) interpersonal skills of teamwork and communication; and (4) conceiving, designing, implementing, and operating systems in the enterprise and societal context. The syllabus is updated continuously. The CDIO initiative is thereby very systematic in describing the standards, explaining the rationale behind the standard, and being explicit about the intended knowledge, skills, and attitudes by formulating a detailed syllabus as well as a rubric with six levels for each standard (see CDIO, 2026). However, what might be most important is that CDIO is an organized network, with conference and knowledge sharing at a high level. There are procedures for becoming an institutional member (CDIO, 2026).

Kristina Edström and I made a comparison between PBL and CDIO and concluded that they were very much the same ideas and principles, although PBL was more focused on learning in the curriculum, whereas CDIO was focused on the design of the curriculum (Edström & Kolmos, 2014). We concluded that these two approaches corresponded and that the more theoretical understanding of PBL could easily complement a curriculum design. The primary distinction between PBL and CDIO lies in the organization of the network. The CDIO project operates as an organization with formal memberships. The PBL societies are organized as open networks, allowing participants to derive inspiration (Kolmos, 2013).

5.2.2 Challenge-Based Learning: The CBL Approach

Challenge-based learning (CBL) does not always imply a systems approach, but it has gained popularity during the last 15 years. The characterization principles are fairly like those of the CDIO and PBL approaches, with the exception that CBL emphasizes the integration of sustainability issues. In the CDIO framework sustainability is built into the standards, whereas in PBL it is up to the students to discover situations that may involve sustainability. The newly published (2025) book, *A practical guide to understanding and implementing challenge-based learning* by Kenan Dikilitaş, offers historical perspectives and cases for how CBL should be understood and how it can be applied in the classroom (Dikilitaş et al., 2025). In this book, it is underlined that CBL is a pedagogy in formation and development, and the authors refer back to the roots of Apple, which emphasize a collaborative learning environment where the content should address the 12 embedded principles by Nichols (Dikilitaş et al., 2025; Nichols & Cator, 2008): That learning should be flexible, scalable, free and open, learner driven, authentic, relevant, technology-integrated, learner-centered, formative, reflective, global, and twenty-first-century skills oriented.

CBL stands on the shoulders of PBL—both the case-based version and the project-based version of PBL. It draws on the same learning theories and origins. It also emphasizes collaborative learning, although project are not mentioned (Nichols

& Cator, 2008). In another guide from KTH, Sweden, the project is core in CBL together with the sustainability issues, which are characteristic for the approach to challenges (Magnell & Högfeldt, 2015). In a study from Eindhoven University, CBL projects are analyzed to identify the project phases and activities undertaken by the students, which clearly indicate the application of projects in CBL (Martin & Bombaerts, 2024).

Another study from Eindhoven University is a review of the implementation of CBL at both a course and a project level that indicates a variation in the way it is implemented and carried out in the learning process. There is an emphasis on the openness of the problems and relationship to sustainability, although with a great variation in the learning outcomes ranging from knowledge to generic competencies (Doulougeri et al., 2024). In this study they also conclude a definition of CBL based on the review, which summarizes the existing practices:

> CBL is an educational approach in which students engage with real-world and multidisciplinary sociotechnical challenges that are ill-structured and open-ended, lacking a single "right" solution. These challenges are ideally presented by societal or industry stakeholders and require the proposal or development of a solution. Acquisition, integration, transfer, and application of knowledge from various disciplines, along with transversal skills such as critical thinking and problem-solving, are needed to develop the solution. The learning process in CBL is active, self-directed, and collaborative, encouraging students to become comfortable with uncertainty. (Doulougeri et al., 2024, p. 21)

Sometimes it might be very difficult to see the differences between PBL and CBL as the above definition could also be valid for PBL. There might be a difference in the scope of the problem, as CBL requires the problem to be ill-structured and real-world, whereas PBL in practice does have a broader approach to also include disciplinary problems. Furthermore, in the CBL emerging traditions, there are no requirements for a project organization of the learning, and therefore the organization of the learning can be much more flexible. In general the CBL should be approached with a variation and flexibility (van den Beemt et al., 2023).

I think from an historical point of view that CBL gained attention in the Netherlands as there was a wish for another profile compared to the case-based PBL at Maastricht University. From an educational transformation perspective, it may be beneficial to rename and rearrange the learning principles occasionally.

No matter CDIO, CBL, or PBL, the collaborative aspect is core, together with the self-directed learning approach. That also counts for other teaching and learning methods such as design-based learning, which is used in design classes and well researched and developed (Atman et al., 2025; Scalone et al., 2019). Understanding collaborative learning is complicated, as all learning is basically individual but will be formed in collaborative spaces and activities. There will be interaction between the inner individual knowledge and social relations, and for learning and in building up resilience this interaction will be crucial. In the following sections I will dig into the understanding of learning.

5.3 Textbooks and Research on Deep Learning

One of the biggest concerns—or at least critical arguments—for academic staff for changing from a traditional curriculum to a PBL has always been the selection and coverage of the content in the curriculum. The fact that students are spending less time in the lecture-based classroom and more time on their projects has resulted in doubt and skepticism as to whether the curriculum could educate engineers with the relevant knowledge and competencies needed.

Indeed, this relies on the assumption that what is told is learned and that the textbook coverage is the core of the curriculum. Both assumptions are widespread in the educational system, and I do not argue against textbooks, but I do indicate that they might be covering only part of the relevant technical knowledge. Studies indicate that the role of textbooks is decreasing in schools and has a different use in engineering education (Lee et al., 2013; Roldán Vera, 2018). In student-centered learning in particular, textbooks are more used as a source of information than as a provider of an overview of a field, and thus the function of textbooks has changed (Lee et al., 2013).

As already indicated, students learn methods and methodologies, and that deep learning in the projects can compensate for the lack of knowledge from taught courses as the students learn to transfer knowledge, methods, and generic competencies from one project to the other. In textbook learning you might achieve the same level of learning outcomes. By using projects, you take some of the time students would use on lectures and apply it to project work instead, creating knowledge gaps. The reasons that this works are twofold: partly that the textbook never covers the entire subject, and partly because one can train students to learn new knowledge areas by the embedded methods of learning in PBL. Furthermore, part of the supervision is also to train students' abilities to transfer methods and knowledge from one problem to another, which then strengthens their abilities to learn.

Research shows that deep learning seems to work within PBL and active learning in general. During the 1980s and 1990s, several research groups within higher education began to work with the concepts of surface and deep learning as a new approach in understanding teaching and learning. Surface learning is where students learn to memorize the text whereas deep learning involves an intention of understanding the text (Biggs, 1988). Understanding is to be able to link and combine new knowledge into original structures of knowledge, which also involves a long-term memory of knowledge (Liu et al., 2022).

Deep and surface knowledge are not only cognitive concepts but also involve a cognitive-affective personality system (CAPS) theory, as students in their deep learning processes for meaningfulness also involve emotional processing. If the learning does not make sense emotionally, it will complicate the cognitive learning (Liu et al., 2022). The CAPS theory basically underpins that both cognitive and affective factors influence our behavior, which is one of several theories for the development of personality, identity and learning, underpinning the importance of both emotional and cognitive factors (Mischel & Shoda, 1995).

Reviewing the literature on PBL and deep learning, there are two types of topics that arise: design for PBL, and the effect of PBL on deep learning. For the design articles, there is an intention to aim at deep learning by active learning, and e.g. Kember shares an institutional approach to promote student-centered learning by sharing good practice from award-winning academics (Kember, 2009).

Project-based learning is related to the concept of usable knowledge, as the application of knowledge is a method for deeper learning. When you have been actively involved in applying knowledge, you will increase understanding and long-term memory (Miller & Krajcik, 2019). Safitri (2024) has conducted a meta-analysis of PBL and transformation at a more general level addressing both scientific and generic competencies (in the article referred to as soft skills) and the study indicates that students attain increased knowledge and competencies (Safitri, 2024).

Sustainable consulting projects are used to develop a deep learning framework that embraces students, institutions, and external partners where students are working together with external stakeholders. Experiences with more than 85 projects indicates that this type of model is successful and that students gain knowledge beyond the expected learning outcomes (O'Brien & Sarkis, 2014). A study from South Africa shows that first-year geography students' approach to deep learning is increased in a two year experimental case study, suggesting that the implementation of PBL helps to increase deep learning (Golightly & Raath, 2015).

The issue with these types of studies is that you never really know what kind of PBL or active learning practices the authors are studying. Even if there are descriptions of the implementations, it is hard to see what the experienced curriculum really looks like for the students. Did they receive a lot of guidance? Were the problem and project well defined or basically open to the students to formulate?

There are also several studies of interest that do not necessarily provide evidence for PBL. The first study encompasses a meta-analysis on the effects of PBL. This review indicates that a curriculum has a positive effect on skills development. In relation to knowledge, the conclusion shows that students increase their learning of knowledge during the first two years with the effect vanishing after that; however, it seems that the students have an increased retention of the acquired knowledge, although in a more narrow knowledge base (Dochy et al., 2003). This article does not indicate anything about deep learning, but the fact that the students remember better what they learned indicates a level of understanding.

This is also underpinned by a study of medical graduates from a PBL institution and a non-PBL institution. The results clearly indicate that the skills and competency part score higher among PBL graduates and they remember the knowledge (Schmidt et al., 2006). We found the same in a longitudinal study on engineering students moving into work. They retain knowledge better from the projects and scored higher on competencies (Kolmos et al., 2021). The above does not show anything directly about deep learning, but the fact that there are significantly better results for skills, competencies, and retaining knowledge might also be worth keeping in mind.

In review study, there is a focus on factors that may encourage or discourage a deep learning approach in student-centered learning environments. The conclusion in this review is mixed and not univocal. There are disciplinary differences; teachers who play an important role, students' satisfaction with the course, intrinsic motivation, self-efficacy, and students who are positive to active learning methodologies all seem to be factors influencing a deeper learning approach (Baeten et al., 2010). Baeten has developed a follow-up empirical study on this topic. None of the empirical studies indicated an increase in students' deep learning approaches, and one of the recommendations from this study is to work on a gradually structured curriculum (Baeten et al., 2013).

Another empirical study followed a smaller group of students to analyze their development of deep learning. This study found that the individual factors such as invested time, study skills and learning skills had a higher impact on the development of deep learning compared to the perceived teaching environment (Postareff et al., 2015). In a study led by Diana Dolmans, 21 studies were reviewed with 11 showing a positive effect, 4 indicating a decrease in deep learning, and 6 studies showing no effect. It was concluded that PBL does seem to have a positive effect on deep learning and a negative effect on development of surface learning. Furthermore, there is a positive effect on students' intrinsic motivation, which is a foundation for deep learning (Dolmans et al., 2016).

The overall picture shows that PBL in some contexts has a positive effect on skills and competency development, retention of knowledge, and slightly positive effect on deep learning.

5.4 Developing Mastery

Whether we are devoted to PBL, CBL, CDIO, or any other student-centered system, we are dealing with collaborative learning and the assumption that students can become drivers of their own learning processes and have agency to lead and influence their own lives. This involves the learner having the experience of owning their own learning process and the possibility to make serious decisions.

This is what was called participatory learning in the Danish origins of the PBL pedagogy and what is called self-directed learning in the origins of the case-based PBL models at the universities of Maastricht and McMaster. That framing, structures, and culture allow you to make choices, and that you have the capacity and belief that you are able to make choices might influence your self-confidence in a positive direction.

This is in the same vein as agency, which is the question of autonomy and ability to make meaningful choices and take ownership of one's own learning and action, or the development of a growth mindset, which is a concept developed by Carol Dweck. A growth mindset is the attitude and belief that one can develop one's own talent or intellectual ability (Dweck, 2016; Yeager & Dweck, 2020). The opposite of a growth mindset is a fixed mindset, where the learner does not believe in the capability to change circumstances or own reactions. A growth mindset is based on curiosity and an open mind, which can be learned by practice several times. Practice will create experiences, which slowly become a habit.

An innovation mindset is very much the same, and a mindset can be seen as dynamic competency development as motivation, learning, and mindsets are core components in competency development and the learning process (Engelsberger et al., 2022). A growth mindset can include the ability to take risks and not be afraid of failure, which can be one of the core elements in resilience and living-with chaos. A study on engineering students identifies growth mindset as a core component in students' persistence when they face difficulties (Campbell et al., 2021).

A growth mindset relates to a feeling of being in charge or mastering one's own learning. During the 1960s and 1970s, there was a trend in education of mastery learning, which is based on Bloom's theories with a stepwise and scaffolded learning process of identifying students' prerequisite knowledge and expectations, integrating formative assessment and feedback from teachers, and often using small group exercises and self-paced learning (Block & Burns, 1976; Bloom, 1956; Guskey, 2010).

Mastery learning is not the same as master/apprenticeship learning, although there are elements that are similar. The master apprenticeship approach is based on a hierarchy where the apprentice learns from the master. Embedded in this understanding of learning is the practice and copying function in learning. The master/apprenticeship approach is arguably the most ancient type of structured education; nevertheless, to learn by copying practice is still a significant way of learning within Western societies.

In its original form, the master/apprenticeship principles are an expression of a practicum, in this context expressed as a trade or a craft. The craft possesses a culture and a set of traditions that define it. As master/teacher, principles rely on practical instruction, the apprentice is instructed in the specified culture and customs, encompassing both verbal and non-verbal expressions (Kvale, 1995; Wackerhausen, 2009). The master/apprenticeship principle has experienced modifications, and recent illumination on the subject has only sparked interest in fragments of it or in its reinterpreted form, as articulated by Schön's reflective practitioners, among others (Schön, 1984).

In a recently published book by Tony Wagner, together with Ulrik Juul Christensen, they look at mastery as a learning strategy in a world of distraction (Wagner & Christensen, 2025). As the title indicates, they see a world of distraction that calls for deeper learning, and their concept of mastery stands on the shoulders of concepts like agency and growth mindsets. In relation to the Bloom-inspired concept of mastery learning and the ancient master/apprenticeship relation, this book represents a renewal of the concept.

The authors emphasize the need for deeper learning in the meaning of critical thinking, creativity, and problem-solving skills, as today's challenge is not to access information but to know how to analyze and apply it in a critical way. Fake news is a daily event, fake information and knowledge are the same, and the ability to critically assess the AI-generated information is urgent. Deeper learning is seen as critical and creative, enacting complex problem-solving by collaborating with others to address real-world problems and transfer or transform that knowledge towards new problems.

Wagner and Christensen do not mention PBL, only different types of project learnings. What is important for them is not to point at a specific teaching and learning methodology but at learning environments that encourage curiosity, risk-taking, reflections, deep learning and align the assessment with deeper learning goals instead of memorizing goals.

Neither the master/apprenticeship approach and mastery learning give explanations to the collaborative aspects of learning, even if mastery learning emphasizes the importance of peer learning. It stresses the importance of the practice dimension, although in university learning it is not enough to have only practice learning.

5.5 Learning Theories Reflecting Collaborative Learning and PBL

Learning theories describe how and why learning occurs. Many learning theories address the same phenomenon: what characterizes students' learning, to understand how we may motivate students to learn better, deeper, and more effectively. I have no plans to write another textbook on learning, and there are many resources available to provide an overview of the landscape of learning theories. However, I have searched for a better understanding of collaborative learning.

Early PBL writings drew on Dewey, Killpatrick, and Freire, as well as Negt, Kluge, Ziehe, and Habermas's critical German theories in the Danish context. During the 1980s and 1990s, many Danish research articles focused on either the educational political system, which was the actual discourse to defend the new reform universities (Borgnakke, 2011; Rasmussen, 1996), or tools and methods for PBL (Algreen-Ussing & Fruensgaard, 1992). None of these early scholars actually focused on learning theories, collaborative learning, or what happens when students are working in teams. During the 1990s, a better knowledge of learning processes emerged, and learning theories came into focus.

Wim Gijselaer was among the first to relate PBL to the learning theories and generate some core learning principles beyond the practice. He formulated three principles based on a range of learning theories that would saturate the practices, or that the practices should take these principles into account (Gijselaers, 1996):

- Learning is a constructive, not a receptive process, which emphasizes active learning.
- Social and contextual factors influence learning, which emphasizes the interaction among learning and its relation to society.
- Knowing about knowing, which is at the meta-level, as well as learning about learning is meta-learning.

Active learning will always be seen as a construction, as the students learn by doing, reflecting, and knowing. The social and contextual factors will influence the students and what they learn. Finally, the meta-level is quite important and something that has been overlooked in the development of PBL; this points to the transformation of learning from one situation to the next; see more about this in Chap. 8.

There is no single learning theory that adequately explains PBL learning processes since they incorporate so many variables in a learning environment. We need learning theories that take into account both individual and social interaction among individuals, as well as interaction between academia and society. Explaining the social relationships between persons in a collaborative process is a challenging task. What knowledge, experiences, personal histories, and social relationships do each individual bring to the table, and how does this translate into collaborative acquisition of knowledge and competencies? Research on student teams clearly reveals that trust among team members is one of the essential elements for team success (Spliid, 2011).

The cognitive and emotional elements are also essential for comprehending the collaboration processes of project teams. The cognitive factor pertains to the acquisition of scientific knowledge; nevertheless, in the absence of motivation, interest, social commitment, and engagement, the impetus for learning is significantly reduced. The cognitive factor must be integrated with the emotional or affective. This was the learning from the early days in my career during the 1980s and 1990s in the Girls and Science and Technology (GASAT) network where we tried to understand the reason why women did not want to go into engineering (GASAT, 1983; Kahle et al., 1987). One of the explanations was that the emotional and social part was lacking in the prevailing curriculum at that time and that one of the solutions would be to integrate more social learning such as PBL (Du & Kolmos, 2009).

5.6 Collaborative Learning in the Classroom

Collaborative learning is an umbrella term for collaborative learning activities in the classroom that won popularity during the 1970s and 1980s (Smith & MacGregor, 1992; van der Linden et al., 2000). There is a conceptual development from cooperative learning to collaborative learning. Cooperative learning is most often understood as the instructional use of small groups in the classroom with the goal of supporting student learning (Johnson & Johnson, 1999).

Collaborative learning is much more seen as a participatory and student-driven learning process, which has more equal power relations between students and teachers (Bruffee, 1995; Dillenbourg, 1999). These definitions are related to the teaching and learning regime and do not really indicate a theoretical understanding of collaboration.

Collaboration takes place in society, networks, and large and small groups or teams. A group is a number of individuals interacting together, whereas a team is characterized by a process in which individuals collaborate for achieving a shared goal. Katzenbach and Smith describe the characteristics for a group versus the progression of teams and have made their observations in companies (Katzenbach & Smith, 2015). Team success is therefore combined with performance and the types are: pseudo team, potential teams, real teams, and high performing teams (Katzenbach & Smith, 2008, 2015). To set up contrasts between groups and pseudo teams on the one side, and potential, real or high-performing teams on the other, it includes variables such as (Katzenbach & Smith, 2008):

- individual and mutual accountability
- identification with the purpose of the team
- individual and collective contributions
- openness and interactive discussions
- shared experiences and commitment
- shared knowledge construction.

These variables can be approached with either a cognitive learning framework, which basically is an understanding of individuals exchanging and sharing knowledge, or a social participatory framework, which is an understanding of participatory construction of knowledge where the individuals form a unified and shared learning system (van der Linden et al., 2000). The first is based on cognitive learning theories and the second on social psychology and situated learning. We contribute to and are affected by our surroundings at the same time.

Social psychology has contributed to understanding collaborative learning and there are different models such as the storming, forming, norming, and performing model, where the conflicts are mentioned as a development stage in the development of the teams (Tuckman, 1965). There is also the approach of defining roles according to positions in the group or social distance (Sjølund, 1975). These approaches lack the learning perspective, and you find the same with Belbin's roles.

Belbin's roles such as the completer, implementer, team worker, specialist, monitor, coordinator, plant, shaper, and investigator can be characterized as roles and functions in a team (Aritzeta et al., 2007; Belbin & Brown, 2022). To have a role can be experienced as something external and a skill or a competency you need to carry out in a collaborative setting and are necessary to learn and perform.

5.7 Holistic View of Learning Theories

Within the field of learning theories, Illeris has made a substantial contribution. As one of the fathers of the Danish PBL traditions, he has had a much more comprehensive understanding of learning in mind with the individual/society and the cognitive/affective (Illeris, 2004).

This has led to Fig. 5.2, where there is a core interaction between individual and social/collaborative on the one side and the content/cognition and incentives/

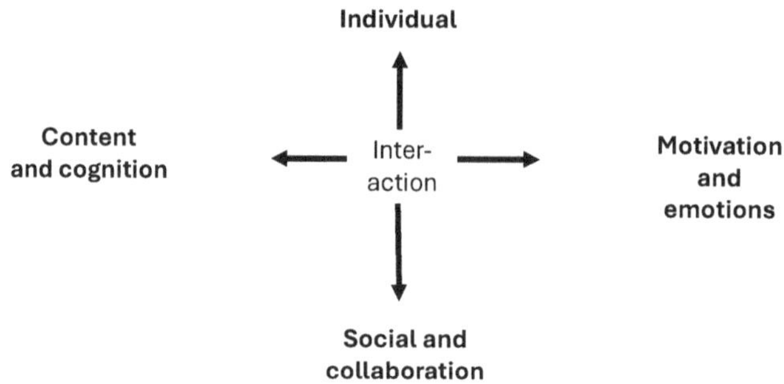

Fig. 5.2 Dimensions in learning theories

emotions on the other. These two dimensions might embrace a PBL practice; however, to cover all aspects is not possible in one theory. We have to combine theories from different fields such as cognitive individual learning, social community learning, psychology, and social psychology.

Student motivation encompasses a wide range of areas, including psychology, cognitive processes, and the affective domain. Understanding student motivation is essential for student-centered pedagogies, as motivation can be influenced by both extrinsic and intrinsic factors. Extrinsic triggers often encompass challenges and social interactions, while intrinsic motivators include personal experiences, emotions, trust, and curiosity. We might not need new learning theories to comprehend the learning that occurs in PBL, but we could require a blend of learning theories to understand some of the mechanisms.

Illeris created a landscape of learning theories and identified three elements of learning: content, motivation, and engagement in and with society (Illeris, 2010). What may be missing from his triangle perspective is social psychology rather than just socialization theories, which typically reduce the individual to a product of social structures. The emphasis in social psychology is on the interaction processes between the individual and the structures and cultures, so the individual exists in the context with their own free will.

The primary perspective of PBL has been to focus on the team as a whole; however, during the last 15 years, there has been a much higher emphasis on the individual as a learner in collaborative contexts. This is a challenge not only for PBL practice but also for our theoretical approach to understanding interpersonal interactions in terms of affective and emotional relationships, as well as cognitive conversation. We need to learn more about how individual and collaborative learning interact.

From the perspectives of Illeris's mapping of learning theories, the emotional part plays a core role in learning. Although we never plan a curriculum from the logic of emotional growth, the learning environment and the students' learning processes are highly affected.

5.7.1 Experiential Learning

As most PBL researchers start with the cognitive perspectives, I would like to take the same starting perspective, but from an experiential point of view. During the 1980s in particular, experiential learning became popular as Kolb had 're-invented' the learning cycle with experience, reflection, conceptualization, and experimentation (Kolb, 1984). These four elements are in a circle going from concrete experience, reflective observation, abstract conceptualization, and active experimentation. Learning might not follow the circle but does jump from experience to experimentation to conceptualization. The point here is more that these are important stages in learning.

It might not be possible to conceptualize anything from one learning event. There must be more learning experiences from which the learner may uncover the pattern; hence, one round among the elements will never lead to a level of conceptualization but to a level of hypothesis that then must be tested. Learning is an iterative process, and students rarely get full comprehension the first time they solve a mathematical equation, but by trying multiple times the understanding develops.

The same goes with learning from experiences in collaborative procedures, e.g. how to communicate and interact with the other team members or how to progress objectives and activities. One plan seems suitable in week 2 but will be reviewed and adjusted in week 4. The process comprises multiple iterations.

Experiences trigger motivation for and are foundations for learning, but experiences alone do not create learning and need interaction with reflection and theory. Reflection, thinking, doing by trying out experiments—iterations are important elements in learning and can facilitate students in planning their own learning in the project teams. They need to dig into the books and listen to the taught courses in order to get the theories, to apply their knowledge, and to create experiences in the learning process. A study on design finds that the experts are doing more iterations in the design process than novices indicating that the mature learner allows constant reflection and questioning in the process (Atman et al., 2008).

The inner experiences will always be a mix of cognitive and emotional parts. In education, we might ask students to do cognitive reflection, and the students will respond to this. But we only get half the story as we do not get their emotional adaptation of the learning. I will go more into reflection in Chap. 6.

5.7.2 Transformative Learning and Identity Development

Transformative learning is a theoretical framework created by Jack Mezirow and other scholars to describe the process of deep personal change that occurs when people critically reflect on their assumptions, beliefs, and values (Mezirow, 2000a). It contributes to collaborative learning while emphasizing that transformative learning entails a shift in viewpoint, which can lead to a change in behavior and a more empathetic and inclusive worldview, and therefore it is basic that the critical reflection leads to some kind of action. Illeris takes this one step further by redefining transformative learning as change in identity (Illeris, 2014).

The learning process has been built on experiences and data from adult learning. The process is characterized by starting with a disorienting dilemma, which is an experience or something that cannot be explained within one's belief and knowledge. Self-examination is the second phase, involving one's own beliefs and understandings for moving to critical assessment of existing assumptions behind one's beliefs. After these first phases, the planning of new learning can take place.

Transformative learning is frequently characterized as a process of unlearning and relearning by critical reflection. It entails challenging firmly held beliefs and assumptions, recognizing biases and limitations, and openness to alternative viewpoints. This is basically understanding learning as personal growth, which is a much deeper level of learning than the cognitive learning taking place in engineering education. Mezirow has been criticized for not including the emotional part and being too focused on the individual. The emotional part can be discussed as Mezirow defines transformation theory like this:

> Transformation Theory's focus is on how we learn to negotiate and act on our own purposes, values, feelings, and meanings rather than those we have uncritically assimilated from others—to gain greater control over our lives as socially responsible, clear-thinking decision-makers. As such, it has particular relevance for learning in contemporary societies that share democratic values. (Mezirow, 2000b, p. 8)

However, Mezirow has a focus on the individual and the interaction with the social dimension and asks that we should be more aware of how we create knowledge and develop our life perspectives. I think that transformative learning is important in understanding student learning both in PBL and non-PBL curricula. The forming of identity and values is often overlooked in engineering education. In a comparison of humanities and science, Barnet indicated that the importance of 'the being', or the identity and values, is much lower in science and engineering, whereas activity has a bigger priority (Barnett & Coate, 2004). Values form part of an identity, which is important in all aspects of life.

Identity formation in a PBL curriculum is complex as it involves both the understanding of the relation between academia and society by the problem identification and the interaction between the individual and collective processes. In teams, transformative dialogues can take place among the participants where there are deep conversations. Similar phases, as described for the transformative learning process, can happen within project teams. The disorienting dilemma, which is the trigger for learning at the deep personal level, might exist if the students start with their own learning experiences. In my own master's thesis, we started with personal reflections on our choice of education, which led to an understanding of moving from lower social classes, becoming a first-generation academic, and understanding new professional functions and roles (Christensen & Kolmos, 1982).

5.7.3 Organizational Learning

Kolb focuses on the individual, yet a team will share techniques and generate common experiences. As a result, team learning will be both individual and collaborative with each team member contributing to shared understandings and conceptualizations. Nancy Dixon contributes to collaborative learning and draws influence from Argyris, Mezirow, and the Lewin/Kolb Learning Cycle. She applies individual learning to an organizational learning viewpoint (Dixon, 2017).

Dixon also runs a blog where she emphasizes conversation and dialogue as means for establishing collective learning and meaning-making, which is highly inspirational (Dixon, 2025). Organizational learning aims to integrate learning processes into organizational understanding, and collaborative learning is one component of that. Her concepts are not intended for project teams, but they can be scaled down to teams of fewer than 10 or 8 people.

Dixon emphasizes meaning structures and distinguishes three categories: individual, accessible, and collective meaning structures. Collective learning is founded on individual learning, and each learner creates his or her own meaning structure. Meaning structures are established by prior learning and experiences, and as Piaget would argue, they have schemas that may be developed or modified through assimilation (adjustment in an existing schema) or accommodation (change of schema) (Piaget, 2013). In Dixon's perspective, meaning might allude to a variety of personal difficulties that the individual prefers not to share.

Accessible meaning is what the individual shares with others, and it is accessible in the sense that it is visible and audible to others. Collective meaning is what all members share in the present and throughout history. Dixon emphasizes dialogue, equality, participation, different perspectives, and shared experiences, all of which may be captured by transparency, user control, and conversation.

Thus, I will add to Katzenbach and Smith's list of success requirements for high-performing teams:

- equal participation
- user-control
- social trust
- transparency
- conversations and dialogues
- space for collective meaning making

Dixon acknowledges that the emotional part does exist in the individual meaning structures and can be an underlying tacit factor, which can be hard to get a hold on. She is also pointing at the collective meaning structures, which can hinder new developments, and which have been internalized into the individual meaning. This is another contribution to the understanding of collaborative learning in the project—to facilitate the student's creation of collaborative meaning structures and let them be aware of the interactive learning processes between the individual and the collective.

5.7.4 Community of Practice

I cannot talk about collaborative learning without community of practices, and much has been written on this topic. I will draw out some core points. Wenger broadens the scope of collaborative learning, and his contribution is that learning is a social activity. Community of practice emphasizes that people learn best when they participate with others in shared, meaningful activities. He distinguishes four kinds of learning (Wenger, 1999):

- Learning as belonging
- Learning as becoming
- Learning as experience
- Learning as doing

He emphasizes that learning is socially integrated and that it is more than simply theory; it is what takes place in real life. It is critical for the learner to be close to the center of a community, and newcomers will learn by observation before gradually participating as full members over time. This is also a master/apprenticeship process, in which individual learners progress from peripheral to full participation.

However, there is also a peer learning aspect in the teams, where the students teach each other depending on how they have shared the work. Lave and Wenger point out that in the old form of master training there is a societal socialization and continuation of traditions, which not only affects the individual craft but has a much broader interface with cultural and societal socialization (Lave & Wenger, 1999). It is the learning of practice by looking at how the master or peers are doing it.

Engineering education is a highly theoretical education, but there are elements of practice in the labs and the projects. You see what others are doing and how they are doing it, not necessarily to copy exactly the same, but to get inspiration for how you can internalize this into your practice. To learn from practice or to learn from seeing what others are doing are overlooked dimensions in pedagogy for higher education in general but also for engineering education in particular. In a community of practice there is a core and a periphery, which can be seen in the light of master/apprenticeship relations.

Learning and identity have an unbreakable connection in Wenger's view. Wenger focuses on the requirements for building communities, which can be official but are mostly informal. Collaboration transforms people into members, and identity, knowledge, and collaboration are inseparably linked. A community can be large or small and collaboration occurs across teams or boundaries. That also means that the project practices will create the students' identities and influence how they interact in future project relations. This is also a way to describe what can happen in a project team where some students might be more in the core and it is a process of creating a common identity.

Collaborative learning is a complicated process. These theorists have contributed various elements to collaborative learning, such as shared experiences, critical reflections and dialogues, collaborative identities, conversations to share knowledge

Table 5.2 Contributions to collaborative learning

Author	Name of theory	Learning	Concepts	Contribution to collaborative learning
David Kolb	Experiential Learning	Learning from experience and theory	Learning cycle: concrete experience, reflective observation, abstract conceptualization, active experimentation	Enhances collaboration through group experience and reflection
Jack Mezirow	Transformative Learning Theory	Individual transformation	Disorienting dilemma, critical reflection, rational discourse, perspective shift	Transformative dialogue
Nancy Dixon	Organizational learning	Knowledge sharing in organizations	Tacit vs. explicit knowledge, learning conversations, knowledge flows	Promotes collaboration through conversational knowledge sharing
Etienne Wenger	Communities of Practice (CoP)	Social, participatory learning	Mutual engagement, learning by belonging, becoming, experience and doing	Embeds learning in collaborative practice communities

and create collaborative meanings, and framing the collaboration through communities where the learning can be characterized as belonging, becoming, experiencing, and doing. These are presented in Table 5.2.

5.7.5 Tacit Collaborative Knowledge

Both Dixon and Mezirow talk about tacit knowledge, a concept described by Polanyi and Sen as opposite to explicit rational knowledge (Polanyi & Sen, 2009). It can be non-articulated knowledge or embodied knowledge that can only be represented via behavior. These two perspectives do not conflict. Knowledge can become embodied; it can be articulated at times or in situations where it is triggered, but not always. Polanyi suggested that scientists should understand that not all information is propositional and in logical order, but that a great deal of our learning and thoughts emerge from this messy and unordered embodied knowledge, which can barely be expressed in words and sometimes in action.

Tacit and nonverbal communication may present both opportunities and challenges for learning through practice. Tacit knowledge has the potential to be a source of intuition for both individuals and teams. In group creativity research, it is well recognized that ideas can emerge through a series of smaller encounters and iterations similar to design processes. One member of the team can submit an idea that prompts comments from the other members, resulting in a continuous brainstorm of associations. The process of contact shapes the team members' culture, which may often remain tacit and develop implicit patterns of interaction. Often in the literature on tacit knowledge and creativity, the jazz band is described as an example where jamming often characterizes the music with intuition and tacit collaboration. I am not sure this is the best example as music is a language; however, the jazz example is good for intuitive collaboration (Sawyer, 2005).

The drawback of tacit knowledge for both individuals and teams might be the difficulty in transferring or transforming information, knowledge, and competencies from one domain to another without articulation. Both the individual and the team must be capable of articulating, communicating, and conceptualizing.

As a result, the process of developing attention and awareness through reflection on these practice experiences will be a crucial component in learning generic competencies. Dixon indicates that tacit knowledge can be triggered and transferred in organizations (Dixon, 2025), and she is correct as our society otherwise would not be able to function, and it might be more difficult for learning to happen. In teams, learning occurs by questioning each other, trying out new angles, reflecting, getting new perspectives, looking at things, etc.

Tacit knowledge is an underlying wealth of ideas, intuitions, emotions, understandings, etc. and suddenly during a discussion things fall into place. In the InterPBL project described in Chap. 7, our discussions clearly illustrate a level of tacit knowledge as a source for new understandings of interdisciplinarity. We test, we draw on the whiteboard, argue, lean back, reflect, try out new formulations, and get a response; a new model is drawn, one that might be deleted during the next meeting as the individual reflections in the meantime bring new ideas. This is an iterative and collaborative process where the ideas are formulated in the collective space but reflected and internalized individually.

Dreyfus and Dreyfus see tacit knowledge, or what they call intuition, as the highest level of mastering expertise. This is where people have gone far beyond rules and procedures and just know what to do in certain situations (Dreyfus, 2004; Dreyfus & Dreyfus, 1980).

Students think that they understand each other after longer periods of discussion; however, they have different comprehensions and this is hard to reveal as only words, connections and body language can be means for communication. As Philippe Baumard points out in his book *Tacit knowledge in organizations*, there is both tacit knowledge at the individual and the collaborative team level (Baumard, 1999). Culture has a tacit dimension, and it can be hard to explain what it means to be Danish or Swedish.

In my early studies, I followed a group of students from their first year to final year and interviewed teams of students about collaboration. There was a clear progression from the first year where the students explained they worked by rules and procedures until the last year when they found it difficult to explain how they collaborated. However, during observation I saw them sitting in pairs in front of computer screens, discussing and writing together (Kolmos, 1999). There was a tacit collaborative learning where the students could finish each other's sentences; however, they could not explicate what was happening.

Whatever the explicit and tacit knowledge, there will be disagreements and conflicts in collaborative learning. How we handle and deal with conflicts has cultural and individual dimensions. It is a natural part of collaborative learning, as there will be disagreements that normally can be argued on a rational and cognitive level. What is more complicated is when disagreements turn out to become conflicts that involve emotional reactions and, indeed, tacit knowledge. Students learn many ways of solving conflicts: a hot chair meeting every week, agreeing after periods of arguing, agreeing that there are personal disagreements and trying to use them constructively, and dissolving the team as the most extreme way.

5.8 Variation in Learning

Variation theory is used in science education to enhance learning, identify differences, and promote reflection, making tacit knowledge more explicit. Phenomenography, inspired the concept of variation in learning, has been used in educational models and research (Marton & Booth, 1997). Learning is defined as a change in perception, experience or understanding and refers to the ability to apply knowledge to specific tasks. Variation theory recognizes individuals' unique perspectives and seeks to comprehend the varying experiences of reality. According to Marton and Booth (1997), learning occurs when students can be aware of several facets of a phenomenon. Comparing perspectives can lead to a more comprehensive understanding of reality and open new possibilities for the future.

There are several perspectives in this chapter, and I have summed up some of the characteristics that derive from the various theories as in Fig. 5.3. I have not given emphasis to any one of them and they are represented as equal variables. They can be regarded as complementary but also as different aspects of collaborative learning. Some of them overlap and some are more abstract than others, but they illustrate some of the elements occurring in a collaborative process.

There is a variation in the contributions to understand collaborative learning and as already stated, there is not one theory to characterize or understand what is going on among participants in a team and the relation to the cognitive learning of problem analyses and problem solving.

Variation among individuals is inherent in collaborative processes, but it can also manifest itself in collective experiences, such as when students experiment with different approaches to share work and they compare and reflect on the experiences. In a comparison process, learners must use abstraction, which involves comparing prior learning experiences with more recent learning experiences. During the abstraction process, learners may be able to generalize their experiences and formulate them conceptually. That happens at a meta-level as either meta-learning or meta-reflection, which involves a critical examination and reflection on one's own interaction in the collaborative learning process to uncover individual and collaborative learning trends. It is a type of continuous action that compares one learning

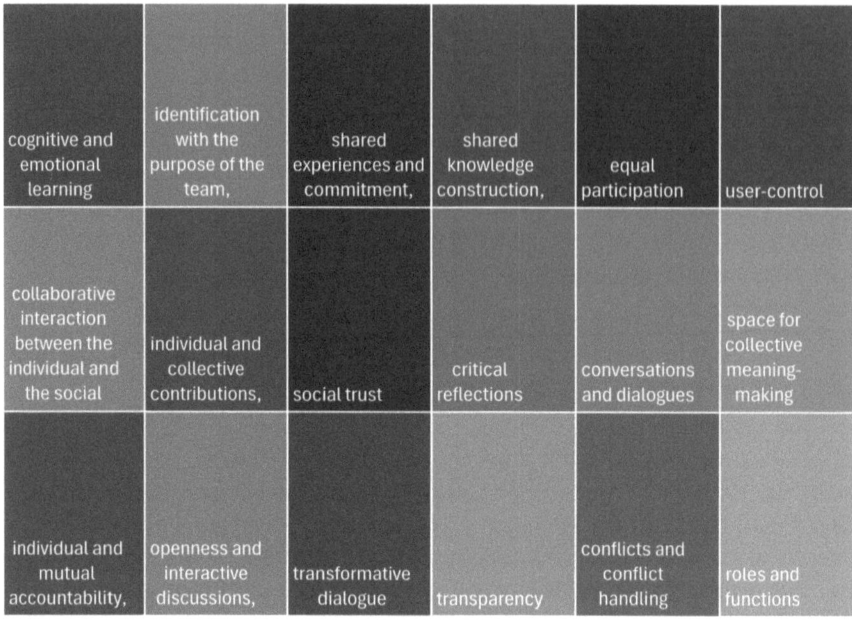

Fig. 5.3 Overview of elements in collaborative learning

experience to another. I will delve further into this in the following chapter on generic competencies and reflection.

5.9 Reflection 5: Living-With Chaos

When we look at the PBL learning principles, we can see that the emotional component and resilience are not always included in the understanding of the learning process. Collaboration involves emotions, and the learning theories discussed consider emotions as a crucial ingredient for motivation. This might be one aspect of reinventing PBL to include a broader goal than scientific knowledge and competencies, such as becoming a reflective citizen, which is more than just branding but an integrated reflective process in the curriculum. Collaborative learning is a prerequisite for more active learning approaches and mastering both one's personal learning and collaborative learning will be a critical component in learning to live with chaos. The development of confidence and self-efficacy is influenced by process of mastery and a sense of being 'on top of things'. No one can handle chaos on his or her own and learning collaborative mastery might be essential in the future.

According to Mezirow, learning derives from disorienting dilemmas, followed by critical reflection, which are steps in a change of perspectives and attitudes. When combined with PBL, this allows students to work on open problems and experience frustration and disorienting perspectives on problems and potential solutions. Students must learn to become holistic decision-makers capable of making decisions collaboratively and across disciplinary and sectoral boundaries. They must learn 'how to swim in the deep water' as part of their resilience development, which entails dealing with ill-structured or complex problems while considering what could go wrong and how prepared they are for such situations.

With learning theories as a foundation, I believe we have many more opportunities to incorporate complex and chaotic problems into the students' learning journeys. Regardless of whether we have PBL or a traditional system, it is crucial that there are elements of complex and chaotic problems and space to reflect this both in professional handling and personal development.

An advantage in a PBL system is that there are good conditions for learning generic competencies, since the students are involved in collaborative learning and working with authentic problems. However, this advantage will include a learning environment which facilitate the reflection of experiences and that is not necessarily the case with PBL practices. I will delve into this in the next chapter.

References

Algreen-Ussing, H., & Fruensgaard, N. O. (1992). *Metode i projektarbejdet: problemorientering og gruppearbejde* (3. opl. ed.). Aalborg Universitetsforlag.

Aritzeta, A., Swailes, S., & Senior, B. (2007). Belbin's team role model: Development, validity and applications for team building. *Journal of Management Studies, 44*(1), 96–118.

Atman, C., Kilgore, D., & McKenna, A. (2008). Characterizing design learning: A mixed-methods study of engineering designers' use of language. *Journal of Engineering Education, 97*, 309–326. https://doi.org/10.1002/j.2168-9830.2008.tb00981.x

Atman, C. J., Zhang, E., Flores, Y., Barar, S. G., & Turns, J. (2025). Good designers do "X": Embracing the breadth of design practices.

Baeten, M., Kyndt, E., Struyven, K., & Dochy, F. (2010). Using student-centred learning environments to stimulate deep approaches to learning: Factors encouraging or discouraging their effectiveness. *Educational Research Review, 5*(3), 243–260.

Baeten, M., Struyven, K., & Dochy, F. (2013). Student-centred teaching methods: Can they optimise students' approaches to learning in professional higher education? *Studies in Educational Evaluation, 39*(1), 14–22.

Bankel, J., Berggren, K.-F., Blom, K., Crawley, E. F., Wiklund, I., & Östlund, S. (2003). The CDIO syllabus: A comparative study of expected student proficiency. *European Journal of Engineering Education, 28*(3), 297–315.

Barnett, R., & Coate, K. (2004). *Engaging the curriculum in higher education* (1st ed.). Open University Press.

Barrows, H. S. (1996). Problem-based learning in medicine and beyond: A brief overview. *New Directions for Teaching and Learning, 1996*, 3–12. https://doi.org/10.1002/tl.37219966804

Baumard, P. (1999). *Tacit knowledge in organizations*. Sage.

Belbin, R. M., & Brown, V. (2022). *Team roles at work*. Routledge.

Berggren, K.-F., Brodeur, D., Crawley, E. F., Ingemarsson, I., Litant, W. T., Malmqvist, J., & Östlund, S. (2003). CDIO: An international initiative for reforming engineering education. *World Transactions on Engineering and Technology Education, 2*(1), 49–52.

Biggs, J. (1988). The role of metacognition in enhancing learning. *Australian Journal of Education, 32*(2), 127–138.

Block, J. H., & Burns, R. B. (1976). Mastery learning. *Review of Research in Education, 4*, 3–49.

Bloom, B. S. (Ed.). (1956). *Taxonomy of educational objectives: The classification of educational goals*. Susan Fauer Company.

Borgnakke, K. (2011). *Et universitet er et sted, der forsker i alt–undtagen i sig selv og sin egen virksomhed. Rapport om den forskningsfaglige baggrund for udvikling af universitetspædagogisk forskning*. Det Humanistiske Fakultet. Københavns Universitet.

Bruffee, K. A. (1995). Sharing our toys: Cooperative learning versus collaborative learning. *Change: The Magazine of Higher Learning, 27*, 12–18. https://doi.org/10.1080/00091383.1995.9937722

Campbell, A. L., Direito, I., & Mokhithi, M. (2021). Developing growth mindsets in engineering students: A systematic literature review of interventions. *European Journal of Engineering Education, 46*(4), 503–527.

CDIO. (2026). CDIO: Conceive, Design, Implement, Operate. Retrieved 23-04-2026 from https://cdio.org/framework-benefits/cdio-syllabus.

Christensen, A.-D., & Kolmos, A. (1982). *Kvindevalg: modsætninger og muligheder*. Aalborg University.

Crawley, E. F., Malmqvist, J., Östlund, S., & Brodeur, D. R. (2010). *Rethinking engineering education: The CDIO approach*. Springer.

Crawley, E. F., Malmqvist, J., Östlund, S., Brodeur, D. R., & Edström, K. (2014a). The CDIO approach. In *Rethinking engineering education* (pp. 11–45). Springer.

Crawley, E. F., Malmqvist, J., Östlund, S., Brodeur, D. R., & Edström, K. (2014b). Student learning assessment. In *Rethinking engineering education* (pp. 165–180). Springer.

de Graaff, E., & Kolmos, A. (2003). Characteristics of problem-based learning. *International Journal of Engineering Education, 19*(5), 657–662.

de Graaff, E., & Kolmos, A. (2007). *Management of change: Implementation of problem-based and project-based learning in engineering*. Sense Publishers.

Dikilitaş, K., Marshall, T., & Shahverdi, M. (2025). *A practical guide to understanding and implementing challenge-based learning*. Springer Nature.

Dillenbourg, P. (1999). *What do you mean by collaborative learning?* Citeseer.

Dixon, N. M. (2017). *The organizational learning cycle: How we can learn collectively*. Routledge.

Dixon, N. M. (2025). *Conversation matters*. https://www.nancydixonblog.com/2009/03/welcome-to-conversation-matters.html

Dochy, F., Segers, M., Van den Bossche, P., & Gijbels, D. (2003). Effects of problem-based learning: A meta-analysis. *Learning and Instruction, 13*(5), 533–568.

Dolmans, D. H., Loyens, S. M., Marcq, H., & Gijbels, D. (2016). Deep and surface learning in problem-based learning: A review of the literature. *Advances in Health Sciences Education, 21*, 1087–1112.

Doulougeri, K., Vermunt, J. D., Bombaerts, G., & Bots, M. (2024). Challenge-based learning implementation in engineering education: A systematic literature review. *Journal of Engineering Education*.

Dreyfus, S. E. (2004). The five-stage model of adult skill acquisition. *Bulletin of Science, Technology & Society, 24*(3), 177–181.

Dreyfus, S. E., & Dreyfus, H. L. (1980). A five-stage model of the mental activities involved in directed skill acquisition.

Du, X., & Kolmos, A. (2009). Increasing the diversity of engineering education – A gender analysis in a PBL context. *European Journal of Engineering Education, 34*(5), 425–437. https://doi.org/10.1080/03043790903137577

Dweck, C. (2016). What having a "growth mindset" actually means. *Harvard Business Review, 13*(2), 2–5.

Edström, K. (2017). *Exploring the dual nature of engineering education: Opportunities and challenges in integrating the academic and professional aspects in the curriculum.* KTH Royal Institute of Technology.

Edström, K., & Kolmos, A. (2014). PBL and CDIO: Complementary models for engineering education development. *European Journal of Engineering Education, 39*(5), 539–555. https://doi.org/10.1080/03043797.2014.895703

Engelsberger, A., Halvorsen, B., Cavanagh, J., & Bartram, T. (2022). Human resources management and open innovation: The role of open innovation mindset. *Asia Pacific Journal of Human Resources, 60*(1), 194–215.

GASAT. (1983). *Contributions to the second GASAT conference.* Institute of Physics.

Gijselaers, W. H. (1996). Connecting problem-based practices with educational theory. *New Directions for Teaching and Learning*, 13–22.

Golightly, A., & Raath, S. (2015). Problem-based learning to foster deep learning in preservice geography teacher education. *Journal of Geography, 114*(2), 58–68.

Guskey, T. R. (2010). Lessons of mastery learning. *Educational Leadership, 68*(2), 52–57.

Hmelo-Silver, C. E. (2004). Problem-based learning: What and how do students learn? *Educational Psychology Review, 16*, 235–266.

Illeris, K. (2004). *Three dimensions of learning: Contemporary learning theory in the tension field between the cognitive, the emotional and the social.* Krieger Publishing Company.

Illeris, K. (2010). *Contemporary theories of learning: Learning theorists in their own words* (reprinted. ed.). Routledge.

Illeris, K. (2014). Transformative learning re-defined: As changes in elements of the identity. *International Journal of Lifelong Education, 33*(5), 573–586.

Johnson, D. W., & Johnson, R. T. (1999). Making cooperative learning work. *Theory Into Practice, 38*(2), 67–73. https://doi.org/10.1080/00405849909543834

Kahle, J. B., Daniels, J. Z., & Harding, J. (1987). *Proceedings fourth GASAT conference, July 24–29, 1987.* University of Michigan.

Katzenbach, J. R., & Smith, D. K. (2008). *The discipline of teams.* Harvard Business Press.

Katzenbach, J. R., & Smith, D. K. (2015). *The wisdom of teams: Creating the high-performance organization.* Harvard Business Review Press.

Kember, D. (2009). Promoting student-centred forms of learning across an entire university. *Higher Education, 58*, 1–13.

Kolb, D. (1984). *Experiential learning: Experience as the source of learning and development* (1st ed.). Prentice Hall.

Kolmos, A. (1999). Progression of collaborative skills. In J. W. A. Conway (Ed.), *Themes and variations in PBL, Vol. 1: Refereed Proceedings of the 1999 Biannual PBL, Conference, July 7–10, 1999, Montreal, Canada.*

Kolmos, A. (2013). Problem- and project-based learning in a global perspective. In A. Aarup Jensen (Ed.), *Visions, challenges and strategies* (pp. 47–66). Aalborg Universitetsforlag.

Kolmos, A., Holgaard, J. E., & Clausen, N. R. (2021). Progression of student self-assessed learning outcomes in systemic PBL. *European Journal of Engineering Education, 46*(1), 67–89.

Kvale, S. (1995). *An educational rehabilitation of apprenticeship learning? Draft.*

Lave, J., & Wenger, E. (1999). *Situated learning: Legitimate peripheral participation* (reprinted. ed.). Cambridge University Press.

Lee, C. S., McNeill, N. J., Douglas, E. P., Koro-Ljungberg, M. E., & Therriault, D. J. (2013). Indispensable resource? A phenomenological study of textbook use in engineering problem solving. *Journal of Engineering Education, 102*(2), 269–288.

Liu, E., Zhao, J., & Sofeia, N. (2022). Students' entire deep learning personality model and perceived teachers' emotional support. *Frontiers in Psychology, 12*, 793548.

Magnell, M., & Högfeldt, A.-K. (2015). *Guide to challenge driven education.* KTH Royal Institute of Technology.

Martin, D. A., & Bombaerts, G. (2024). What is the structure of a challenge based learning project? A shortitudinal trajectory analysis of student process behaviours in an interdisciplinary engineering course. *European Journal of Engineering Education*, 1–31.
Marton, F., & Booth, S. (1997). *Learning and awareness*. Lawrence Erlbaum Associates.
Mezirow, J. (2000a). *Learning as transformation: Critical perspectives on a theory in progress* (The Jossey-Bass higher and adult education series). ERIC.
Mezirow, J. (2000b). Learning to think like an adult. In *Learning as transformation: Critical perspectives on a theory in progress* (pp. 3–33). Jossey-Bass.
Miller, E. C., & Krajcik, J. S. (2019). Promoting deep learning through project-based learning: A design problem. *Disciplinary and Interdisciplinary Science Education Research, 1*(1), 1–10.
Mischel, W., & Shoda, Y. (1995). A cognitive-affective system theory of personality: Reconceptualizing situations, dispositions, dynamics, and invariance in personality structure. *Psychological Review, 102*(2), 246.
Nichols, M., & Cator, K. (2008). *Challenge based learning white paper*. Apple.
O'Brien, W., & Sarkis, J. (2014). The potential of community-based sustainability projects for deep learning initiatives. *Journal of Cleaner Production, 62*, 48–61.
Piaget, J. (2013). *Origin of intelligence in the child: Selected works* (Vol. 3). Routledge.
Polanyi, M., & Sen, A. (2009). *The tacit dimension*. University of Chicago press.
Postareff, L., Parpala, A., & Lindblom-Ylänne, S. (2015). Factors contributing to changes in a deep approach to learning in different learning environments. *Learning Environments Research, 18*, 315–333.
Rasmussen, P. (1996). Projektarbejde og reformuniversiteterne. In F. Kupferberg (Ed.), *Kreativt kaos i projektarbejdet*. Aalborg Universitetsforlag.
Roldán Vera, E. (2018). Textbooks and education. In *The Palgrave handbook of textbook studies* (pp. 103–114). Springer Nature.
Safitri, I. (2024). Transformation of student science achievement through project-based learning (PjBL): A meta analysis study. *Scientiae Educatia: Jurnal Pendidikan Sains, 13*(2).
Sawyer, R. K. (2005). *Social emergence: Societies as complex systems*. Cambridge University Press.
Scalone, G., Joya, A. J., Shroyer, K. E., & Atman, C. J. (2019). Design intentions: Engineering students looking ahead to their future design behavior. In *2019 ASEE Annual Conference & Exposition*.
Schmidt, H. G., Vermeulen, L., & Van der Molen, H. T. (2006). Longterm effects of problem-based learning: A comparison of competencies acquired by graduates of a problem based and a conventional medical school. *Medical Education, 40*(6), 562–567.
Schön, D. A. (1984). *The reflective practitioner: How professionals think in action* (1st ed.). Basic Books.
Sjølund, A. (1975). *Gruppepsykologi*. Gyldendal.
Smith, B. L., & MacGregor, J. T. (1992). What is collaborative learning. In *Towards the virtual university: International online learning perspectives* (pp. 217–232). Emerald.
Smith, K., Maynard, N., Berry, A., Stephenson, T., Spiteri, T., Corrigan, D., Mansfield, J., Ellerton, P., & Smith, T. (2022). Principles of problem-based learning (PBL) in STEM education: Using expert wisdom and research to frame educational practice. *Education Sciences, 12*(10), 728.
Spliid, C. M. (2011). *Mastering projects and processes in the Aalborg PBL model. PBL across the disciplines: Research into best practice* (pp. 555–568). Aalborg Universitetsforlag.
Tuckman, B. W. (1965). Developmental sequence in small groups. *Psychological Bulletin, 63*(6), 384.
van den Beemt, A., van de Watering, G., & Bots, M. (2023). Conceptualising variety in challenge-based learning in higher education: The CBL-compass. *European Journal of Engineering Education, 48*(1), 24–41.
van der Linden, J., Erkens, G., Schmidt, H., & Renshaw, P. (2000). Collaborative learning. In *New learning* (pp. 37–54). Springer.
Wackerhausen, S. (2009). Collaboration, professional identity and reflection across boundaries. *Journal of Interprofessional Care, 23*(5), 455–473.

Wagner, T., & Christensen, U. J. (2025). *Mastery why deeper learning is essential in an age of distraction*. Basic Books.

Wenger, E. (1999). *Communities of practice: Learning, meaning, and identity*. Cambridge University Press.

Yeager, D. S., & Dweck, C. S. (2020). What can be learned from growth mindset controversies? *American Psychologist, 75*(9), 1269.

Chapter 6
Generic Competencies, Reflection and Experimentation

6.1 The Transition from Education to Work

Since the 1990s there has been a focus on skills and competencies. In higher education there has been a trend to move to outcome-based and competency-based education, carried forward by policy agencies such as the EU Bologna process (Malhotra et al., 2023). The primary goal of engineering education is to produce engineers who not only understand science but also possess the skills and competencies required to apply theoretical knowledge to innovation and technological development. As universities transitioned from elite to mass universities, this development was critical for the legitimacy of higher education in general, and particularly engineering education.

During the 1960s, engineering colleges at the bachelor level were established with a stronger link to practice; the TEKNIKUM in Denmark and the Fachhochschülen in Germany are just two examples of these trends to bridge the gap between the theoretical scientific lenses and engineering practice. In the 1990s, however, the competency perspective required the integration of application, practice, and societal usefulness in higher education in general. Teamwork, collaboration, and project management are among the criteria used by engineering education accreditation boards. Politically, employability was cited as one of the primary reasons for promoting a competency and outcome agenda in higher education.

Employability was elevated to the political agenda, with studies emphasizing competencies such as communication, management, flexibility, and teamwork (Baytiyeh & Naja, 2012; Ford et al., 2019; Stiwne & Jungert, 2010). Other studies point out that the most important skills and competencies besides the technical knowledge are critical thinking, problem solving, and innovation, which are among the highest-ranked competences among academic staff (Magnell & Geschwind, 2019; Magnell & Kolmos, 2016).

A. Kolmos, *Engineering Education for Interdisciplinary Futures*, Emergent Directions in Engineering Education 2,
https://doi.org/10.1007/978-3-032-32540-2_6

Young and Chapman (2010) provide an analysis of the identified generic competencies necessary for career transitions, which also indicates the necessary competencies learned in engineering education. They reviewed frameworks from Western countries such as the EU, the United States, Canada, Australia, and New Zealand. In their summary, they identified 58 skills divided into six categories: basic, conceptual, personal, people, business, and other (Young & Chapman, 2010). This is consistent with the findings of numerous other studies on employability and issues in the transition from education to work, which have identified similar discrepancies. Several studies on graduate engineers' experiences and responses to their education identify gaps in competency profiles (Collet et al., 2015). Communication is seen as most important for engineers to function in a job together with interprofessional collaboration (Baytiyeh & Naja, 2012; Collet et al., 2015; Gaughan et al., 2017). Studies even emphasize that the most prominent changes will be in the first three months, adjusting to teamwork, communication, and types of self-directed learning (Ford et al., 2019).

A Swedish study revealed that successful functioning in a job requires generic and metacognitive skills and an ability to adjust to a (new) context (Nilsson, 2010). Results from a Danish study, PROCEED-2-WORK, which followed a cohort of Danish students from all Danish institutions from their first year of study to their first job, indicated that around one third of engineering graduates do not feel prepared for competencies such as social responsibility, environmental impact, ethics, management skills, business knowledge, and leadership (Kolmos & Holgaard, 2019; Kolmos et al., 2021).

The US academic pathway project was a longitudinal study on graduates with one of the largest cohorts of 57 young engineers interviewed in 2007. It focused on what knowledge they apply in their first job and where they learned it (Brunhaver et al., 2017). The study indicates a clear trend that knowledge was learned at university, although a high percentage still indicated that they also learned on the job, whereas professional skills and organizational knowledge and skills were primarily learned on the job. After four years, this group of engineers was asked again, and the results indicate that technical knowledge seems to be less important compared to the professional skills, which increase in importance.

Lutz and Paretti (2021) analyzed the transition from a perspective of shifting contexts or organizations, giving interesting insights into culture and social approaches to learning, as generic competencies are learned in a context, and therefore it might still be hard to transform the generic competencies learned in a university context to a work context. Every organization will be different, with various goals and cultures, so it might be challenging to adopt and understand the values (Lutz & Paretti, 2021). Within the same vein, another US study on early career identity clearly underpins that identity is formed by both work and non-related work activities (Huff et al., 2019).

The transition from education to work involves an increasingly long list of competencies that students need to learn to be employable. The list of relevant competencies is constantly growing, and somehow universities have become a marketplace for competencies. Education cannot respond to such extensive lists of competencies by incorporating everything into the curriculum. Education must emphasize core competencies and prepare students to learn throughout their lives.

Many of the studies also emphasize that beyond technical proficiency, engineers need a set of skills and abilities that transcend disciplinary boundaries, and that learning communicative and collaborative competencies is needed for adaptability and the ability to apply knowledge in multiple contexts (Hall et al., 2019). Disciplinary boundaries are created within academia and do not reflect the boundaries in engineering work. Engineering practice is in many ways to be compared with interdisciplinary or transdisciplinary practices in academia and, in a recent study, the results indicate that young engineers are surprised by the requirements of their first jobs, such as the level of generic skills, working in parallel projects, taking on several different roles, and not at least being highly dependent on colleagues (Friedrichsen et al., 2024).

Competency-based teaching and learning is relatively new in higher education in engineering education. Learning competencies and skills includes practical learning and self-development. Take collaboration as an example: we cannot learn collaboration solely by reading about it; instead, we must practice and learn how to behave in various situations. If our reactions are inappropriate, we must learn to respond differently. Reactions are strongly related to emotions, and developing our reactions will necessitate emotional participation and development. Emotions influence how we see ourselves, our self-understanding, personal values, and identity. Therefore, we not only talk about learning rational and logical knowledge, but also self-development, identity development, and emotional maturity.

PBL is a comprehensive pedagogy that includes collaboration, project and people management, and problem analysis from multiple perspectives. It allows project teams to learn from practice and learn to develop competencies in practice. However, learning is more than just practice. It also involves reflection and more theoretical conceptualization, which education must incorporate into the curriculum as learning outcomes.

6.2 Generic Competencies

Generic competencies refer to what are also known as transferable skills, professional skills, graduate attributes, transversal skills, soft skills, key skills, core skills, twenty-first century skills, employability competencies, and process competencies. These competencies include lifelong learning skills that enable individuals to perform not only in disciplinary and academic domains but also across academic disciplines, work, and social life. For all of them, there is a general focus on enabling students and learners to understand the practice, and contribute with solutions to, and apply, share and contribute their knowledge in communicative and collaborative organizational settings (Male, 2010; Young & Chapman, 2010).

I use the term 'generic competencies' to describe competencies or skills that are applicable across fields. This is not to say that a student who has acquired a generic competency in one subject can immediately apply it in another. In the InterPBL project (see Chaps. 7 and 8), we found that students who learned generic skills or competencies in disciplinary settings struggled to transfer and transform these skills into interdisciplinary contexts. However, it appeared to be easier to transform and apply generic competencies from a disciplinary to a narrow interdisciplinary context compared to a transition from a disciplinary to a broader interdisciplinary context (Kolmos et al., 2024).

Collaboration among individuals or project teams can be challenging due to divergent interests and priorities (see Chap. 5). Research also identifies that even the learning of generic competencies is contextual, and in a study comparing engineering and business students attitudes to the learning of generic competencies, engineering students did not acknowledge these as much as business students (Chan & Fong, 2018). Another study indicates that disciplines and culture have a significant impact on the learning of generic competencies (Jones, 2009).

Generic competencies play an important role in an individual's capacity to make independent choices, take action, and be responsible for the action taken. There is also the concept of agency, which in essence encompasses the individual's empowerment, decision-making, communication, influence, resilience, and collaboration (Eteläpelto et al., 2013). The learning and development of generic competencies are closely intertwined with an individual's sense of agency. These competencies empower individuals to take charge of their lives, make informed choices, navigate challenges, and manage life circumstances.

Handling societal chaos caused by political, economic, health, climate, or technological factors, involves a strong self who can act and handle their own life individually as well as collaboratively. Active learning methodologies reflect agency agendas and might frame the learning environment in which the students learn to act and interact. However, it is not done by just that. It needs to be combined with inner reflection and the buildup of individual and collaborative self-confidence.

6.3 Competencies and Meta-Competencies

I have chosen to apply the concept of competencies rather than skills. There are overlapping understandings of skills, competencies, and competences, with marginal differences in many cases. The definitions of these concepts are not less messy compared to the understanding of PBL. In many of the Anglo-Saxon countries, the term 'skills' is used, whereas most of the EU countries use the term 'competences'. The Australian researchers, Hager and Gonczi, argue that moving from skills to competence or competency is a matter of going from a more specific and narrowly defined ability to a richer and more flexible work performance (Hager & Gonczi, 1996). The competency concept has to be understood as more abstract compared to skills and to encompass a possibility for flexibility.

In an article with exactly the same title as Hager and Gonczi's article, 'What is Competence', Le Diest and Winterton give their suggestion to the understanding of competence (Le Deist & Winterton, 2005). First, they argue that the difference between competence and competency is that competence belongs to a functional area mostly seen in the UK, whereas competency reflects the behavior seen in the US. This usage of the concepts 'competence' and 'competency' is used inconsistently in the literature, which is well documented in this article reviewing the understanding in the US, UK and Continental Europe (Le Deist & Winterton, 2005). In a recently published report, there is a definition of competence as:

> the ability to integrate and apply contextually appropriate knowledge, skills, and psychosocial factors (e.g., beliefs, attitudes, values and motivations) to consistently perform successfully within a specified domain. (Vitello et al., 2021, p. 4)

They talk about knowledge, skills, and psychosocial factors as key elements for performance. In their definition of competence versus competency, they understand competence as attached to a person as a concept to capture the holistic qualities from a person, whereas competency is seen as more narrow and detailed, attached to a task or activity (Vitello et al., 2021). This definition is based on the structure/task versus the qualities the person has developed. Even if this sounds logical, the choice of concepts does not follow logical lines, and I only use competence if it is referred to by an author, which is often the case for European writers; otherwise, I use the competency-competencies concept.

As already mentioned, Young and Chapman (2010) identified a long list of 58 competencies based on the review of the same geographical areas as Le Deist and Winterton and came up with six basic categories: basic, conceptual, personal, people, business, and other (Le Deist & Winterton, 2005; Young & Chapman, 2010). Le Deist and Winterton, based on their review, propose a holistic model of competences to address the ever-changing landscape of this field: cognitive, functional, social, and as an extra layer, meta-competences (Le Deist & Winterton, 2005).

This is a holistic model where the cognitive and functional competencies are the competencies relevant for work, as the social covers the behavior and attitudes. On top of this is the meta-competency, which is the competency to reflect and develop the three other competencies. Le Deist and Winterton underpin that this model is an analytical model, and how the individual learner will experience their competency development will be a much more integrated process.

The learning of competencies is different from theoretical learning. In the following section I will emphasize reflection and experimentation as core elements in the learning of generic competencies.

6.4 Case of Teaching and Learning Generic Competencies at Aalborg University

Almost from the beginning of Aalborg University's history, the first-year curriculum at the Faculty of Engineering and Science contained PBL learning outcomes complementary to generic engineering discipline outcomes formulated as "... *skills in independent problem-formulation and problem-solving"* as well as *"... skills in collaboration and communication"* (Translated) (Den Teknisk Naturvidenskabelige Basisuddannelse, 1974–95).

During the 1970s, the problem-based and project-organized pedagogy was new to students, and they needed to get tools and understanding of new types of learning processes. The first PBL pioneers at Aalborg University wrote textbooks to facilitate the learning in the projects teams (Algreen-Ussing & Fruensgaard, 1992; Algreen-Ussing et al., 1975, 1978). Since the beginning of 1974, this course has had many titles from 'Methods in Project Work', 'Collaboration, Project Management and Learning', and now 'Problem-Based Learning'. The objectives of this course have also undergone change from being an introduction to PBL to an explicit learning of the competencies framed by the PBL environment at Aalborg University. A literature review on PBL competencies indicates that most often, there is a focus on teamwork combined with problem analysis and project management (Boelt et al., 2022).

This is a difficult course to teach because students believe they already know how to interact in project teams, so their first reaction might be that this is just common sense and a waste of time or that it is overly theoretical and useless. However, after attempting to complete a 15 ECTS project once, they begin to realize that there may be a need to learn more about problem analysis, collaboration, project management, leadership, conflict resolution, communication with external stakeholders, interactions with facilitators, and so on.

There is a pattern in how students acknowledge their learning in these courses. Maybe students believe they can do it on their own the first time, but they may encounter difficulties in some areas. If students fail in some areas, they become aware and realize that there is something they can do better. For example, a project team made up mostly of introverted and highly intelligent students who failed to coordinate and meet on a regular basis, would result in a project and report that clearly lacks consistency from problem formulation to conclusion.

The second time they do a project, the students begin to consider how they can benefit from the courses taught in the first semester and start to learn from their mistakes. They create structures and systems for their project process and begin to structure their communication. They go back to textbooks in the field and discuss how they can improve communication and coordination.

The third time, it starts to work for them because they have two experiences and a better understanding of how to manage project learning. Structures, processes, and routines are incorporated to form a type of tacit knowledge or nonverbalized action skills, in which they act without carefully planning or discussing it beforehand or afterwards. Nonverbal conceptual collaboration can be extremely effective and can characterize teams at a high level (Katzenbach & Smith, 2015).

If teams retain the same members, the 'tacit' division of labor will persist. Although tacit collaboration can accommodate expertise that would otherwise be difficult to obtain, knowledge that is not verbally expressed can be difficult to apply in other collaborative situations. This will not be a problem for the experienced expert, who will have a wealth of situations and experiences to draw from. For a student in a learning situation, it is critical that they learn to apply their learning experiences in new situations and articulation of learning is a part of this process.

This course has always served as a learning scaffold for students by helping them understand and analyze problems, organize their learning, and write reports. It has provided a systematic introduction to PBL and what can be learned through participation in such learning processes. This is a unique feature of the Aalborg PBL model. Many PBL programs claim that students learn PBL skills and competencies; however, it is largely up to the students to learn how to navigate these processes.

Internationally, it is reported that professional competencies such as ethics, oral and written communication, teamwork, and project management have increasingly been taught, for example, in the United States since 1994 (Hauhart & Grahe, 2014). In European countries, several centers or courses were established to teach these competencies; however, it was not integrated into the curriculum.

The students' prerequisite has also changed over the years. During the '70 s and '80 s, there was not much PBL in the school system or in high school, and it was necessary to build up the knowledge from the ground. This first phase was very much instructional in the sense that it gave tools and was organized by Helle Algreen-Ussing and her team of colleagues. We have written a chapter on the development of the philosophy of this course in the Wiley Handbook on PBL (Kolmos et al., 2019).

We found three main phases of the course that we have called the instruction, experiential learning, and instrumental learning (see Table 6.1). During the first phase instruction, the scaffolding of PBL was build up by conceptualization and definition of core activities. The core of the courses was very much instruction and only presenting one way of doing things. I am not sure if the students followed this, but they were given advice.

A core driver for the development of the teaching and learning of PBL skills has been to develop meaningfulness in the studies for students. At Aalborg University, if students had project experiences from high school, the students in their first year were surprised by the fact that the projects at Aalborg University were more comprehensive and were semester-long, which meant that they had to collaborate in new ways. They could no longer avoid conflicts and needed to gain an overview of a much bigger subject area. The course, therefore, had to develop and cover topics in new ways such as project management, collaboration, methodology for problem analysis, and conflict management, and reflection was one of the core learning methodologies. Therefore, in the second phase of the course development, we developed reflection and experimentation as core elements in the learning.

6.5 Learning by Reflection and Experimentation

The second phase of the course was the experiential learning of the course, see Table 6.1. Reflection and experimentation were core elements. Reflection might be another difficult concept to grasp. 'Is reflection like soap?' is one of the article titles

Table 6.1 Three phases in teaching the PBL course (Kolmos et al., 2019)

	Learning environment	Learning approach	Learning activities	Assessment
Phase 1: Instruction (1974–1994) Methods in project work	Responsibility: facilitators and course lecturers Learning organized in interdisciplinary theme based 'large groups' Managed by interdisciplinary faculty groups and collaborative planning	Experience-driven instruction Course linked to project and course facilitators Collective focus	Course lectures Co-supervision Project diary Process description PBL skills assessed as part of the group-based project assessment	Written process analyses submitted together with the project and assessment in the oral project assessment
Phase 2: Experiential learning (1994–2006) Collaboration, Learning, and Project Management	Initiatives and responsibility: students, course lecturers and facilitators Learning organized in discipline-based semester groups Managed by disciplinary faculty groups and collaborative planning	Facilitated experiential learning based on evidence from practice Course linked to project and project facilitators Collective focus	Course lectures Co-supervision Structured reflection by analyses of collected data Process analysis PBL skills assessed as part of the group-based project assessment	Written process analyses submitted together with the project and assessment in the oral project assessment
Phase 3: Instrumental learning (2006–2017) PBL course	Responsibility course lecturers and facilitators Learning organized in discipline-based semester groups Managed by discipline-based coordinators	Compelled Instruction Course linked to project Individual and collective focus	Course lectures Consultants Individual written assessment of course Process analysis PBL skills assessed as part of individual or group-based project assessment	Written process analyses submitted together with the project and assessment in the oral project assessment combined with a separate individual assessment

and it demonstrates very well how difficult the concept and activities are to describe and conceptualize, and definitions are either too broad or too narrow (Schaepkens et al., 2022). Furthermore, reflection is difficult to research, despite being said to have a positive effect on reducing burnout and stress, developing professional competencies, and facilitating the transition from one context to another (Schaepkens & Lijster, 2023).

Several studies and research have been conducted in the medical field, where improving practice is an essential component of the profession. This also applies to many other professional fields, such as engineering, though the practice will differ.

The question, however, of what happened and whether something could be done differently applies to all professions.

Reflection is difficult to incorporate into education because it combines inner thoughts with prior knowledge or experiences. Rather than a more authentic reflection of how practices are experienced and felt, education frequently takes an instrumentalist approach by defining steps and procedures (De la Croix & Veen, 2018). De la Croix and Veen (2018) propose three strategies for incorporating reflection into education: (1) abandon the checklist approach; (2) adopt a personal approach to reflection; and (3) focus on improving reflection practices.

This very good advice was not followed in the integration of reflection in engineering education during the 1990s, as reflection was seen as a much more logical and rational process. The competency agenda called for a more systematic development of learning from practice.

During the 1990s, we realized that the instruction-based learning approach needed to change, and we accomplished this by building learning on reflection theories by Cowan, Kolb and Schön (Cowan, 2006; Kolb, 1984; Schön, 1984). Through Fig. 6.1, Cowan introduced reflection as a three-step model: a 'for' pre-reflection, an 'in' reflection, and a retrospective 'on' reflection were used as framework. Schön had already introduced the 'in' and 'on' reflections as an intertwined process for a reflective practitioner (Schön, 1984), and Cowan's argumentation for a before reflection was to create a baseline to which the in and on reflections could be compared. In many ways, this is reasonable and logical argumentation.

Getting students to do a pre-reflection is difficult. Anyone who has tried to get students to set expectations for courses or projects will agree with this statement. As anticipation and foresight studies become more common, the competency of expectation and imagination has emerged as one of the new competencies in the recent SDG literature (Rieckmann, 2017). What we know about the future is that the choices we make today shape the future, and that the choices are influenced by our conscious and unconscious expectations for the future.

Even if foresight and future orientation are recognized as competencies that students must acquire, it is difficult to envision what I, as a student, personally require or desire to learn. A scientific foresight differs from a personally formulated

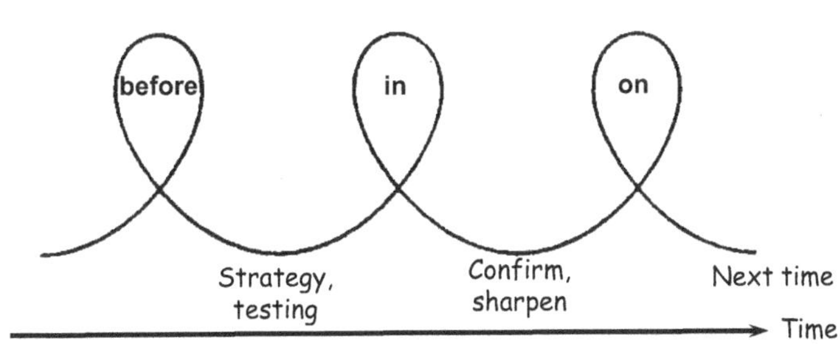

Fig. 6.1 John Cowan's reflection loops (Cowan, 2003)

expectation, and students struggle to identify their personal and academic strengths and weaknesses. As academics we struggle to articulate our expectations, which is fundamentally human. It is difficult to judge, evaluate, or express explicit expectations and self-perceptions.

What distinguishes a PBL-organized education is that the reflection process should occur at both the group and individual levels. Schön describes reflection in and during the process as 'reflection-in-action', which aids in the development of the reflective practitioner. These ongoing reflection processes—or iterative learning—exist in individual project groups, and in the first year of study they are formalized in a final process analysis that includes a retrospective component as a form of reflection-on-action.

6.6 Could We Start with Experiment Instead of Reflection?

In the second phase, reflection methods were integrated into the PBL course, and the pedagogy in the course was changed so the students always had to reflect on their experiences, see Table 6.1.

Asking students to reflect—whether before, during or after—was a challenge! They thought looking backwards was boring; they wanted to get things done, so we concluded that this approach to reflection was ineffective. The more theoretical approach didn't work. The elements of reflection on experiences, conceptualization, and experiments were followed in a linear manner, and we began to experiment with the concept through the Kolb's cycle.

The idea was to take a different approach to learning by allowing students to design their own experiments and reflect on them. We combined Kolb's approach to learning with Schön's approach of 'reflection-in-action' and 'reflection-on-action' (Kolb, 1984; Schön, 1983). 'Reflection-in-action' is a process where reflection and experimentation are integrated and somehow take place at the same time. 'Reflection-on-action' is reflection taking place at a distance, and it is more the process of looking backwards; like the reflection in Kolb's learning circle, see Fig. 6.2.

Using experiments and asking the students to plan and make experiments was much more aligned with their interests and motivations. For engineering students, this created more motivation as they were asked to experiment and be creative with their own collaborative processes (Hansen et al., 2003; Kofoed & Kolmos, 2004).

We discovered three distinct levels of reflection by allowing students to start with activities such as designing their own project management system. If we allow students to plan and conduct experiments without organized reflection, they will develop a common sense that is free of argumentation. It's similar to a trial-and-error approach in that they intuitively test out a variety of ideas, which can be extremely useful for reflective practitioners.

However, in order to communicate with others or apply your experiences and skills in novel situations, you must use a language other than tacit and intuitive acts.

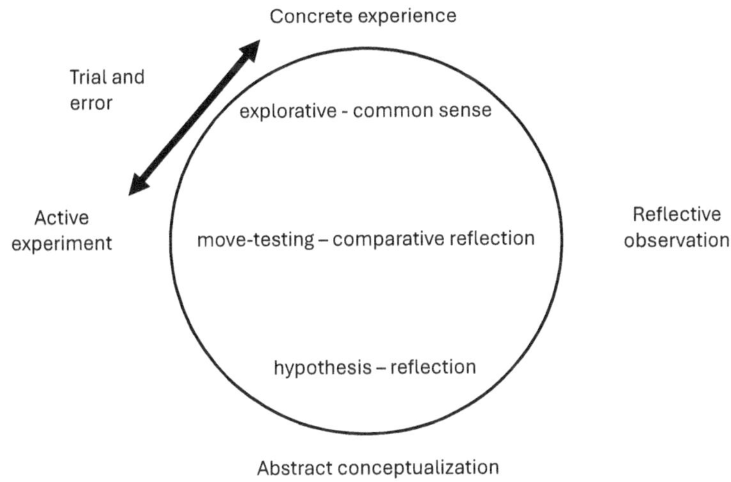

Fig. 6.2 Reflection model for learning PBL competencies (Kolmos et al., 2004)

Schön refers to this as the explorative phase, which occurs "when action is undertaken only to see what follows, without accompanying predictions or expectations" (Schön 1983, p. 145).

After the explorative phase, we saw that the students started to have several experiences and associated iterations. We called this comparative reflection. When students have had multiple experiences, they can begin to compare them. For example, with project management plans, they might begin to analyze different systems and determine which ones are most suited to different contexts and purposes. This is comparable to Schön's move testing notion. The move-testing experiment suggests that there is an aim—an expected outcome of the action, e.g., I do this and that in the group so that we get to know each other better—or that we can create a better social environment.

Finally, there is conceptual reflection, where students compare their understandings to existing concepts or theories to form a more thorough understanding and language. Schön refers to this as hypothesis, and it is much more similar to a regular experiment with specified objectives and a more comprehensive hypothesis comprising of different variables (Schön, 1983).

It is critical to use experiments as a learning strategy. Experiments are designed as a learning method for a variety of reasons:

- Experimentation enables you to clearly define your objectives, techniques, and outcomes. It helps you become more aware of your actions.
- It is a strategy for creating unique experiences that can foster creativity, new thinking, and innovation.
- It is well suited to engineering students' active rather than passive approaches to learning.

- Experimentation has long been used to describe engineering learning and creativity.

The application of this experimentation-reflection method in the course was very successful. The process-analyses that students submit with the project report have significantly improved. After submitting two pages of common-sense reflections during phase one of the course, the students began submitting 10–12 pages of documentation of their collaboration. They examined team functions, how they distributed tasks, how they resolved problems, and, particularly, how their project plans changed over the semester project. The students' willingness to understand more about teamwork, learning, and project management increased, and we were asked at one point to reduce our expectations so that they could focus on technical and scientific learning!

6.7 The Generic PBL Competencies in the Curriculum

Table 6.1 shows that in the third phase the conditions changed more than the learning strategies. Because the learning of PBL methodologies had to be adjusted to new curricula frames, the course had to have an individual course assessment. The individual dimension was therefore introduced. This aligned the PBL course with the other technical and scientific courses, adding prestige to the field.

From 2010 onwards, PBL was institutionalized through accreditation and the teaching of PBL competencies. Responses from external advisory boards to the different study programs came with the same message: "Your students do not have any idea of their own collaborative, communicative and organizational competencies."

Even if learning PBL competencies was a requirement, it was never explicitly included in the curriculum beyond the first semester. The vice-deans of education requested that the UNESCO Centre promoted this new initiative. In less than a year, the concept had been defined, and action plans were in place. Around 2015, the pro-rector initiated a change for the students to reflect their PBL competencies during their entire study. The PBL competencies should be integrated into the formal curriculum and at the UNESCO PBL Centre, we were given the task to identify the specific PBL competencies, develop teaching and learning methodologies, and help the study board integrate it into the curriculum (Holgaard & Kolmos, 2019, 2020, 2021a).

We defined the specific PBL competencies as four major clusters of competencies: understanding of the problem, collaborative aspect, the structural aspect, and at the meta level including reflection across the specific domains (Fig. 6.3).

Figure 6.3 shows the four PBL competency fields specified with examples of practices under each competency. The problem-oriented competencies concern the core learning objectives of analyzing societal and scientific problems, which can be an element in motivation for learning as one identifies a purpose.

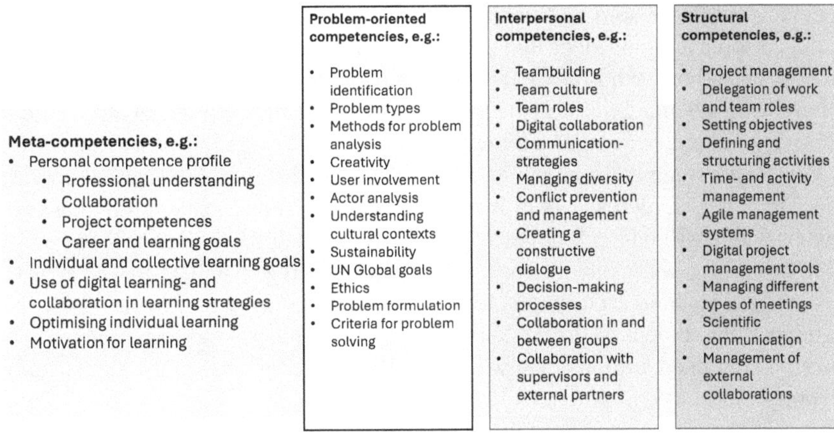

Fig. 6.3 PBL competencies (Holgaard & Kolmos, 2020; Holgaard et al., 2019)

The interpersonal competencies relate to the relationship between the student and other people such as fellow students, supervisors, external collaborators, etc. The structural competencies characterize the relationship between the student and the tools/methods for structuring the problem-based learning process. The interpersonal (between people) and the structural (people in relation to organization) competencies can be summarized in one concept as process competencies which we did in our early writing on this subject, i.e., competencies that are focused on handling the process itself around the problem-oriented work (Kolmos & Kofoed, 2002).

The metacognitive competencies relate to the relationship between the students and the learning process, which supports, connects and creates new thinking in the three competencies above.

Integrating meta-competencies into the curriculum was new, and we defined it as the ability to combine specific competencies, reflect on and develop these.

From the research side, we developed this framework with specific competencies belonging to each cluster. The study boards then integrated selected parts of these in the formal curriculum together with a requirement for students to submit a PBL competency profile. Employability competencies are also one of the strengths of PBL, as students learn to collaborate within their own professional field. They learn to share and structure knowledge from within the professional field, as these are integrated processes (Holgaard & Kolmos, 2020; Holgaard et al., 2019).

The UNESCO PBL Centre was also responsible for running workshops for members of study boards and other interested parties (Holgaard & Kolmos, 2021b). It was the first time we had explicitly attempted to formulate PBL competencies in a progression perspective, and while I would not claim to have arrived at a final conclusion, we had at least begun the process. There is a progression to be developed for each of the specific PBL competencies, which includes both the students' learning outcomes and teaching methodologies.

Collaboration is more than just doing things together; there is a progression in how well we transfer and transform our skills and competencies from one team to the next. It might be hard to be explicit about this, but Dreyfus and Dreyfus have developed some learning steps that might be very useful to frame this process, starting with the novice who needs introductions and rules, moving to an advanced beginner who has limited experiences from various situations, to being a competent and proficient learner who has a more holistic view, and ending up as an expert who has developed routines in adjusting to new and challenging situations (Dreyfus, 2004; Dreyfus & Dreyfus, 1980). It must be understood as a scale more than anything else, as it will be hard to indicate precisely which level one has reached.

We developed a framework for teaching this; see Fig. 6.4. The first year is to transform the students from pupils to students, and the last year is to facilitate the transition from student to professional. During this process, there is a continuous reflection combined with new knowledge and areas, depending on the single study program.

We kept the course during the first-year program, which we had been running for years and combined this with workshops. The intention was to facilitate students in writing a PBL competency profile as the last activity. This was a great idea; however, in practice, it did not really work. Some students took this seriously and were very happy for the possibility and others thought this was nonsense (Boelt et al., 2023).

These workshops, particularly the one on competency profiles, were implemented during the first phase of Covid-19, when everything had to be done remotely for both teachers and students. Those were not ideal conditions. Even if we had a great idea and teaching material to back it up, the conditions influenced our practice. After the first trial, it improved. Today, due to limited resources and a lack of overall success, PBL profiles are no longer required in the curriculum. However, it is available as a free study activity for students who want it.

We created a series of reflective questions to facilitate the development of the PBL profiles (see Table 6.2). The questions are a great support for what one can start to reflect on. We set up questions for each competency area in a logic of:

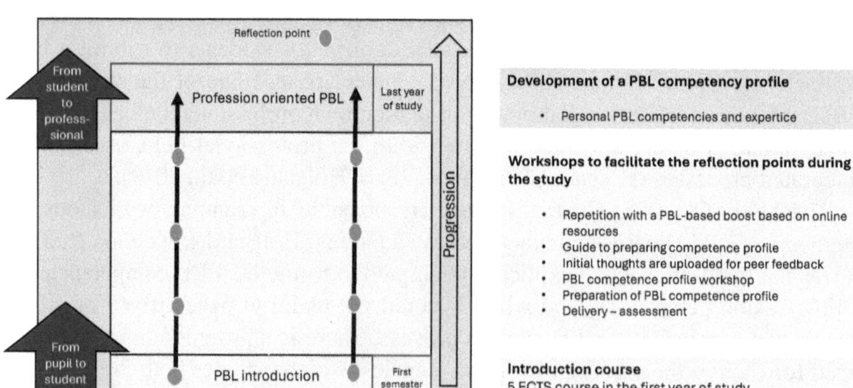

Fig. 6.4 Framework for teaching PBL competencies (Holgaard & Kolmos, 2019)

Table 6.2 Questions to facilitate PBL profiles (Holgaard & Kolmos, 2021b)

	Problem	Structure/ project management	Inter-personal	Meta-reflection
Experience	What types of problems have you worked with in the problem-based project work (concrete/abstract, practical/ theoretical, stable/ dynamic, etc.)? How have you worked with problem solving (specialized vs distributed, sequential vs iterative, operational vs entrepreneurial, etc.)?	How have you organized project work in the teams, that you have been a part of? What tools, including digital tools, have you used to plan your project work? What competences did you gain from working with these tools?	What competencies has the initiating phases of the projects given you in relation to establishing and defining your team (team building, team culture, team roles)? Think about your experiences with communication and collaboration in your project work. What competencies has this collaboration given you in terms of working in teams, across teams and with external partners?	Think about your experiences from preparation of a process analysis. Which concepts/sets of concepts were brought into play to analyze problem design, open problem solving, project design, collaboration and learning? What competences have you gained in this concern to optimize your own learning as well as organizational learning? Have you incorporated reflection on learning styles into your work to balance your own learning style and make synergy of learning styles in teams? What competences has this given you?
Variation	Are your skills primarily aimed at one type of problem (which one, provide examples) or are your skills broadly aimed at several different problem types (provide examples of the variation)?	Have different problem types prompted different project designs? If so, what competencies has this given you in terms of situating your project design?	Which competencies have you gained in working with people from different backgrounds—disciplines, professions, and cultures? What intercultural competencies have you gained from this?	What competencies have you gained in using different reflection approaches to reveal opportunities for change?

(continued)

Table 6.2 (continued)

	Problem	Structure/ project management	Inter-personal	Meta-reflection
Your contribution (strengths/ role/ facilitate others)	What are your strengths in terms of identifying, analyzing, formulating, and solving a problem? What have been your contributions in this process?	What has been your role in project management in the teams you have been in, and what competencies have this given you?	What has been your primary role in team collaboration?	How did you work to develop your problem-based competencies through your studies (e.g., personal learning goals, strategies, use of theory and method, experiments, evaluations, new goals)?
Potentials	How do you think your way of approaching and working with a problem empower you in your future work life?	What experiences has your project work given you in terms of the project manager role, and how will you transfer these experiences to further projects?	What competencies has external collaboration given you when interacting with, and perhaps even being part of, another organization? How will this affect your skills in future business-to-business relations?	What competencies do you have in transferring learning outcomes from one situation to another? Do you expect to use these competencies in your future educational and professional practice?

- Think of own experiences
- Reflect on variations of experiences from similar situations or the same experiences from different situations
- Identify your own contribution in the team
- Think of potentials

We emphasize variation and guide students towards structured reflection on their own experiences using these four angles. Furthermore, the students are motivated to consider their own contributions to the team. Satisfaction with one's contributions and learnings are frequently blurred in collaborative learning because it is difficult to identify each contribution after they have been edited several times. Satisfaction or being content contributes to the development of identity and self-confidence. These questions were formulated in a guide, and we have no idea how many students used them. They do, however, give an idea of how we have tried to facilitate the students' learning of PBL competencies.

Reflecting on the questions today, not many years after they were formulated, the logic in the question, as well as the intention to structure students' reflections, was clearly developed with good intentions and I think that the questions give inspiration to what students can reflect on. My own consideration today is that maybe we were far too structured, and we left out the emotional aspects of learning. We do not request it directly or explicitly. Emotions exist implicitly because experiences are always associated with both cognitive and emotional aspects. However, this is not what we ask students to report on, and the emotional component is hidden or tacit.

6.8 Curriculum Variation in the Learning of PBL Generic Competencies

The curriculum approach is the framing of the learning of PBL competencies. There is a practice and a learning culture. To facilitate the students' learning, there are factors in the curriculum that can be improved, in particular the following:

- The PBL practices that frame the students' experiences.
- How the students reflect on their experiences and develop strategies for improvement.
- How the educational system frames, facilitates and assess the progression of learning.

The interaction of these three aspects can serve as the foundation for learning PBL generic competencies. There is no doubt that the development of competencies in the engineering field is characterized by continuous reflection on experiences, similar to the reflective practitioner. However, if students only learn to do in-reflection, they risk ending up in trial-and-error, as previously mentioned. Learning to be an ongoing reflective practitioner necessitates the presence of structured reflection loops that allow for ongoing reflection on how the individual interacts in new situations. As a result, there is a need for structured reflections across the curriculum.

Structured reflection can take many different forms, as previously stated. Essentially, there is deductive reflection, in which theories, models, and concepts are used to comprehend and analyze practical experiences. There is also the reverse process of inductive reflection, which begins with a comparative analysis of experiences and the formation of understanding as a foundation for understanding theory. The ideal scenario is for both types to be used to connect experiences and theory on a deeper level of understanding.

As mentioned, PBL competencies are types of practice competencies and learning occurs through reflection on practice and conceptualization and analysis. Integrating reflection methods into education is therefore an important element, as students cannot be expected to develop these competencies on their own. With guided reflections on variations of experiences, students can gain deeper insight into their problems, projects, collaboration, and learning competencies.

A condition for this is that there is an intended variation in the curriculum that should be aligned with the learning outcomes. Variation in PBL models contributes to a greater degree of variation and breadth in PBL competencies. If students reflect on the same experiences from participating in two very similar project types, the differences will not be as obvious as if they reflect on, say, the division of labor and collaboration in an interdisciplinary project and a discipline project. The variation will put the students' routines to the test because they cannot always use the same procedures, methods, and forms of collaboration.

Variation is one way to break up routines and tacit knowledge. This applies to variation at all levels in the PBL curriculum, such as described by (Ryberg & Kolmos, 2023):

- Problems from simple to complex
- Project types from narrow discipline projects to complex megaprojects
- Interaction between courses and projects
- Supervision and collaboration forms
- Collaboration with external partners from case to partner
- Different physical facilities
- Different digital systems
- Project scope from a few ECTS to many ECTS
- Project length from shorter to longer courses
- Team size from smaller to larger groups
- Team composition from national to international groups
- Team formation from bottom-up to top-down way of creating teams
- Internal group collaboration forms from sharp division of labor to integrated collaboration

Reflection on variation is an important source for learning professional methods in problem analysis and problem delimitation, as well as process competencies for managing and participating in problem-based project collaboration. Reflecting on variation allows the learner to become aware of the uniqueness of the experience. The typical illustrative example is a color comparison, in which looking at two blue colors next to each other causes you to notice differences, characteristics, and possibilities. The same holds true for project management. If you have only tried one way of collaborating and project management, then that is the experience and knowledge you carry forward, whereas if you have experienced 2 or 3 different ways, then you carry experience and knowledge about different ways of doing it and the ability to be flexible and able to see which processes are important for new tasks.

6.9 Reflection 6: Living-With Chaos

The increasing number of competencies that universities are supposed to address is concerning since it only adds to an overloaded curriculum. Overload might cause shallow learning. Problem solving, teamwork, and management methods are addressed in holistic learning settings like PBL. However, for students' competency development there should be combined practice experiences, reflections, and additional experiments. In our experience with developing reflection methods, students found experimenting to be the most appealing technique since it allows them to customize and shape their practice.

In terms of reacting to chaos, experimentation with practice has the advantage of allowing students to learn more systematically, beginning with actions and then reflecting. Chaos necessitates instant reactions, which can involve rapid adjustments. In this sense, in general learning generic competences will be beneficial to developing chaos reactions.

I have pointed at reflection and concluded that even if reflection can be organized in many ways, it must be structured. At least structured so as to give it space in the curriculum and provide resources for its facilitation. The collaborative learning process is also complicated and drawing on Chap. 5 with respect to learning theories, there are plenty of issues to reflect on in collaborative learning. However, even with structured reflections, reflection will always be experienced as some kind of soup. It is difficult to get hold on.

In engineering education, I have always been cautious about the scientific approach and discourses I presented. From the beginning of my career, I understood the importance of emotions in learning. However, you rarely discuss emotions in engineering education; discourses are rational, logical, factual, and preferred, indicating a quantitative approach. Getting qualitative research methodologies recognized by engineering education research societies has been a long process. Deans want numbers, and numbers can influence educational change decisions. Logical and rational discourses are important, however, given the increasing anxiety among young people, emotional reflection must be articulated. It might help to talk about emotions, and one might discover that they are not alone in the world. Sharing fosters a sense of community and belonging as a caring and helpful culture.

Emotional reflection may be among the strategies for increasing students' resilience. However, it needs re-evaluation and the incorporation of an interdisciplinary curriculum and boundary crossing as important components, as well as a stronger emphasis on resilience that goes beyond the traditional focus on knowledge, skills, and competencies. This differs from the conventional competencies and skills agenda in that it involves emotional and personal development beyond the typical educational context.

For collaboration, humility and modesty are basic aspects and it is possible to reflect on these. The issue is that when goals are formulated in the curriculum it needs to be followed by assessment. And for more personal competencies such as modesty and humility to other people, disciplines or cultures, these might be difficult to assess.

We need to find places where we can share and develop counterstrategies or coping strategies. Doing this individually could create more fear and feelings of loneliness. Being part of a community or a collective culture sharing ideas and concerns might be one of the recipes for dealing with a chaotic world.

References

Algreen-Ussing, H., & Fruensgaard, N. O. (1992). *Metode i projektarbejdet: problemorientering og gruppearbejde* (3. opl. ed.). Aalborg Universitetsforlag.

Algreen-Ussing, H., Skov-Petersen, B., & Fruensgaard, N. (1975). *Rapportskrivning i projektorienteret undervisning*. Aalborg Universitetscenter, den teknisk-naturvidenskabelige basisuddannelse.

Algreen-Ussing, H., Fruensgaard, N. O., & Skov-Petersen, B. (1978). *Projekt og rapport: rapportskrivning i projektorganiseret undervisning*. Aalborg Universitetsforlag.

Baytiyeh, H., & Naja, M. (2012). Identifying the challenging factors in the transition from colleges of engineering to employment. *European Journal of Engineering Education, 37*(1), 3–14.

Boelt, A. M., Kolmos, A., & Holgaard, J. E. (2022). Literature review of students' perceptions of generic competence development in problem-based learning in engineering education. *European Journal of Engineering Education, 47*(6), 1399–1420.

Boelt, A. M., Holgaard, J. E., & Kolmos, A. (2023). A thematic analysis of engineering students' experiences of teamwork in problem-based learning. *International Journal of Engineering Education, 39*(3), 627–642.

Brunhaver, S. R., Korte, R. F., Barley, S. R., & Sheppard, S. D. (2017). Bridging the gaps between engineering education and practice. In *US engineering in a global economy* (pp. 129–163). University of Chicago Press.

Chan, C. K., & Fong, E. T. (2018). Disciplinary differences and implications for the development of generic skills: A study of engineering and business students' perceptions of generic skills. *European Journal of Engineering Education, 43*(6), 927–949.

Collet, C., Hine, D., & Du Plessis, K. (2015). Employability skills: Perspectives from a knowledge-intensive industry. *Education + Training, 57*(5), 532–559.

Cowan, J. (2003). *Curriculum development: A booklet to support a staff development workshop*. Pædagogisk Udviklingscenter.

Cowan, J. (2006). *On becoming an innovative university teacher: Reflection in action* (2nd ed.). Society for Research into Higher Education & Open University Press.

De la Croix, A., & Veen, M. (2018). The reflective zombie: Problematizing the conceptual framework of reflection in medical education. *Perspectives on Medical Education, 7*(6), 394–400.

Den Teknisk Naturvidenskabelige Basisuddannelse. (1974–1995). *Normalplan 1974–95*. Aalborg Universitet.

Dreyfus, S. E. (2004). The five-stage model of adult skill acquisition. *Bulletin of Science, Technology & Society, 24*(3), 177–181.

Dreyfus, S. E., & Dreyfus, H. L. (1980). A five-stage model of the mental activities involved in directed skill acquisition.

Eteläpelto, A., Vähäsantanen, K., Hökkä, P., & Paloniemi, S. (2013). What is agency? Conceptualizing professional agency at work. *Educational Research Review, 10*, 45–65.

Ford, J. D., Paretti, M. C., Kotys-Schwartz, D., Howe, S., Gewirtz, C., Deters, J., Chowdhury, T. M., Ott, R., Alvarez, N. E., & Knight, D. (2019). Transitioning from capstone design courses to workplaces: A study of new engineers' first three months. *International Journal of Engineering Education, 35*(6), 1993–2013.

Friedrichsen, D., Holgaard, J. E., Kolmos, A., & Routhe, H. W. (2024). Surprises at work. A study of early career engineers. In *SEFI 2024 proceedings*. SEFI: European Association for Engineering Education.

Gaughan, K., Craps, S., Pinxten, M., Saunders, G., & Leandro-Cruz, M. (2017). Professional roles and employability of future engineers.

Hager, P., & Gonczi, A. (1996). What is competence? *Medical Teacher, 18*(1), 15–18.

Hall, K. L., Vogel, A. L., & Croyle, R. T. (2019). *Strategies for team science success*. Springer. https://doi.org/10.1007/978-3-030-20992-6

Hansen, S., Kolmos, A., & Kofoed, L. (2003). Teaching and learning process competencies by experimenting and reflecting. *Das Hochschulwesen. Forum für Hochschulforschung, -praxis und -politik, 6*, 235–242.

Hauhart, R. C., & Grahe, J. E. (2014). *Designing and teaching undergraduate capstone courses*. Wiley.

Holgaard, J. E., & Kolmos, A. (2019). Progression in PBL competences. In *SEFI annual conference 2019 European Society for Engineering Education. Annual Conference proceedings*.

Holgaard, J. E., & Kolmos, A. (2020). *Progression in PBL competences*.

Holgaard, J. E., & Kolmos, A. (2021a). *Competence profiles for problem based learning (PBL): Guide for students for preparing a PBL competence profile*. The Aalborg UNESCO Centre. Retrieved May 21, 2026 at https://vbn.aau.dk/da/publications/competence-profiles-for-problem-based-learning-pbl-guide-for-stud/.

Holgaard, J. E., & Kolmos, A. (2021b). *A guide to facilitate development of: Progression of PBL competences*.

Holgaard, J. E., Søndergaard, B. D., & Kolmos, A. (2019). *Guide og katalog til PBL progressive læringsmål: Forberedt for ENG og TECH*.

Huff, J. L., Smith, J. A., Jesiek, B. K., Zoltowski, C. B., & Oakes, W. C. (2019). Identity in engineering adulthood: An interpretative phenomenological analysis of early-career engineers in the United States as they transition to the workplace. *Emerging Adulthood, 7*(6), 451–467.

Jones, A. (2009). Redisciplining generic attributes: The disciplinary context in focus. *Studies in Higher Education, 34*(1), 85–100.

Katzenbach, J. R., & Smith, D. K. (2015). *The wisdom of teams: Creating the high-performance organization*. Harvard Business Review Press.

Kofoed, L., & Kolmos, A. (2004). Development of process competencies by reflection and experimentation. In V. M. S. A. Gil & H. Hooghoff (Eds.), *Challenges in teaching & learning in higher education* (pp. 77–90). University of Aveiro.

Kolb, D. (1984). *Experiential learning: Experience as the source of learning and development* (1st ed.). Prentice Hall.

Kolmos, A., & Holgaard, J. E. (2019). Employability in engineering education: Are engineering students ready for work? In *The engineering-business nexus: Symbiosis, tension and co-evolution* (pp. 499–520). Springer.

Kolmos, A., & Kofoed, L. B. (2002). Developing process competencies in co-operation, learning and project management. In *4th World Conference of ICED*.

Kolmos, A., Fink, F., & Krogh, L. (2004). *The Aalborg PBL model: Progress, diversity & challenges*. Aalborg University Press.

Kolmos, A., Bøgelund, P., & Spliid, C. M. (2019). Learning and assessing problem-based learning at Aalborg University: A case study. In *The Wiley handbook of problem-based learning* (pp. 437–458). Wiley.

Kolmos, A., Holgaard, J. E., & Clausen, N. R. (2021). Progression of student self-assessed learning outcomes in systemic PBL. *European Journal of Engineering Education, 46*(1), 67–89.

Kolmos, A., Holgaard, J. E., Routhe, H. W., Winther, M., & Bertel, L. (2024). Interdisciplinary project types in engineering education. *European Journal of Engineering Education*, 1–26.

Le Deist, F. D., & Winterton, J. (2005). What is competence? *Human Resource Development International, 8*(1), 27–46.

Lutz, B., & Paretti, M. C. (2021). Exploring the social and cultural dimensions of learning for recent engineering graduates during the school-to-work transition. *Engineering Studies, 13*(2), 132–157.

Magnell, M., & Geschwind, L. (2019). A seamless blend of research and professional practice: Dual coupling in engineering education. *Higher Education Research & Development, 38*(4), 807–818.

Magnell, M., & Kolmos, A. (2016). Employability and work-related learning activities in higher education: How strategies differ across academic environments. *Tertiary Education and Management*, 1–12.

Male, S. A. (2010). Generic engineering competencies: A review and modelling approach. *Education Research and Perspectives, 37*(1), 25–51.

Malhotra, R., Massoudi, M., & Jindal, R. (2023). Shifting from traditional engineering education towards competency-based approach: The most recommended approach-review. *Education and Information Technologies, 28*(7), 9081–9111.

Nilsson, S. (2010). Enhancing individual employability: The perspective of engineering graduates. *Education + Training, 52*(6/7), 540–551.

Rieckmann, M. (2017). *Education for sustainable development goals: Learning objectives*. UNESCO.

Ryberg, T., & Kolmos, A. (2023). Hybridisation, variation and reflection: New PBL curriculum principles. In *PBL in a digital age* (pp. 27–38). Aalborg Universitetsforlag.

Schaepkens, S. P., & Lijster, T. (2023). Mind the gap: A philosophical analysis of reflection's many benefits. In *Helping a field see itself* (pp. 117–126). CRC Press.

Schaepkens, S. P., Veen, M., & De La Croix, A. (2022). Is reflection like soap? A critical narrative umbrella review of approaches to reflection in medical education research. *Advances in Health Sciences Education, 27*(2), 537–551.

Schön, D. A. (1983). *The reflective practitioner: How professionals think in action*. Temple Smith.

Schön, D. A. (1984). *The reflective practitioner: How professionals think in action* (1st ed.). Basic Books.

Stiwne, E. E., & Jungert, T. (2010). Engineering students' experiences of transition from study to work. *Journal of Education and Work, 23*(5), 417–437. https://doi.org/10.1080/13639080.2010.515967

Vitello, S., Greatorex, J., & Shaw, S. (2021). *What is competence? A shared interpretation of competence to support teaching, learning and assessment. Research report*. Cambridge University Press & Assessment.

Young, J., & Chapman, E. (2010). Generic competency frameworks: A brief historical overview. *Education Research and Perspectives, 37*(1), 1–24.

Chapter 7
Interdisciplinary Learning Within or Across Programs?

7.1 Introduction

If we are going to address chaos and complexity, one single discipline cannot provide the solutions by itself. Interdisciplinarity is essential and, therefore, interdisciplinary collaboration is a must to learn. This was the message in Chap. 2. PBL was intended to be interdisciplinary, but it was primarily used as a pedagogy for disciplinary learning, as described in Chaps. 3 and 4.

At Aalborg University, the majority of the programs are almost like disciplinary silos, and the PBL practice, which has a global impact, is primarily disciplinary. It has become an inflexible system because the existing curriculum lacks elasticity in terms of electives. An elective system where students can choose their own courses allows for individual specialization, as well as a more interdisciplinary profile. This is the case at Harvard University, for example, with student populations who learn to combine knowledge in their own personal learning track and create their own curriculum (Habbal et al., 2024).

A curriculum structure may be a barrier to the freedom of creating an individualized curriculum (Ryberg & Kolmos, 2023). Although one may indicate that specialization will be included in the chosen projects, projects do not always include open and unstructured problems. Furthermore, the decision to choose problems and thereby projects is made collectively in a project team; although a team consists of individual students who influence the collaborative decisions, the individual student might not win in the negotiation process. This is one of the issues with flexibility in the PBL system.

As a relatively new university in its early stages of development, Aalborg University has a culture of establishing new programs rather than launching cross-disciplinary initiatives. This has also been a part of the reform universities' image to create new disciplines or programs that could fill gaps in the existing university traditions or point in new directions.

A. Kolmos, *Engineering Education for Interdisciplinary Futures*, Emergent Directions in Engineering Education 2,
https://doi.org/10.1007/978-3-032-32540-2_7

Developing new programs takes a great deal of effort and the labor market is a major issue. Where do the candidates find work after their studies? For the university, this was a strategy to attract those students who might not have been interested in a purely technical approach. This was a success because the new programs combined aspects of humanities and social science with technical solutions. Establishing new interdisciplinary programs can also be used to educate 'translators', boundary brokers, boundary workers, and others—those candidates with the ability to bridge gaps between professional fields.

However, a new professional and social understanding emerges in these new interdisciplinary programs. After a few years, an anthropologist will be able to communicate with engineers and vice versa. A new cultural practice will slowly emerge, not necessarily integrated, but respectful, in which two disciplines acknowledge the conditions. In the beginning of such interdisciplinary programs, there might be competition for resources and scientific profiles. The shared curriculum will gain acceptance over time, establishing a new discipline or a new community of practice. There are many battles to fight, and studies indicate it is not only internal in the interdisciplinary program but also complicated. In academia in general, research indicates that interdisciplinary programs are less respected as compared to disciplinary programs (Millar, 2019).

Academics perceive interdisciplinarity in different ways. A newly published study identified three overall approaches to interdisciplinarity each with several sub-themes: capacities, identity, and institutional practice. The capacity approach covers knowledge integration, generic competencies, and epistemology, whereas the second theme contains the tensions between generalists and specialists. The third theme covers the contexts at micro, meso, and macro levels (Ming et al., 2025). What this study reveals is that interdisciplinarity has multiple understandings, and therefore various practices.

Learning interdisciplinary collaboration may require a rethink. If there is really need for an organization of knowledge and new technical-scientific competencies, then a program or a community will be the best solution. If it is a question of students' learning of interdisciplinary competencies, the learning of collaborating across disciplines might then be the solution.

7.2 What is a Discipline?

Before I present my rethinking of interdisciplinary collaboration, I will make a detour on the characteristics of a discipline. Most of the literature on interdisciplinarity jumps directly to interdisciplinary competencies or learning. My latest research project is the InterPBL project funded by the Poul Due Jensen Foundation, also known as the Grundfos Foundation. Our focus is not on interdisciplinary education but on interdisciplinary collaboration and learning in projects as we study types of interdisciplinary projects across existing disciplines (Kolmos et al., 2025). With the argument that new interdisciplinary programs end up as new disciplines, we emphasize it is the crossing of disciplinary boundaries that is the core in interdisciplinary learning. A fundamental question has been: What is a discipline? What must we learn to cross, combine, or integrate?

We cannot talk about interdisciplinarity without having an understanding of a discipline (Chettiparamb, 2007). Disciplines are social constructions that are emphasized by many scholars (Abbott, 2002; Yates et al., 2016). A discipline is a subset of knowledge that is studied and taught by using shared epistemologies and methodologies. A university-based academic discipline is a subset of knowledge studied and taught at that institution.

Disciplines are widely understood to be social creations (Abbott, 2002; Yates, Woelert, Millar, & O'Connor, 2016), and sharing epistemologies and techniques is simply one of the features. Disciplines have been institutionalized by university structures since the nineteenth century, with departments, educational programs, professional groups, journals, conferences, and cultures (Fensham, 2004). Chettiparamb argues that there are three dimensions: The first concerns the scientific approach including branch of knowledge, theories, methods, and ways of knowing, the second is the relation to society, and the third is the institutional approach (Chettiparamb, 2007).

Abbot emphasizes that the disciplines' structures are currently determined by university systems. He specifically identifies well-established research and career structures as hurdles to transdisciplinary futures, as well as cultures that shape habits and practices (Abbott, 2002). However, even if the university and scientific institutions form the disciplines and restrict any change (because this would entail a more general shift in structures and cultures), institutional cultures differ and influence the discipline in the knowledge approach.

7.3 New Interdisciplinary Engineering Disciplines

Interdisciplinarity and new interdisciplinary programs have always existed. Over the last 50 years, new educational programs have emerged that include mechatronics, nanotechnology, biotechnology, global business studies, health technologies, environmental engineering, biomedical engineering and systems engineering. To establish new programs and ultimately a new discipline has not been easy, as most examples require two disciplines to collaborate on common research and educational programs and build up research societies, conferences, journals, and research communities. All of this takes time and in my own area of engineering education research, it has taken at least 30 years to gain recognition. The history of material science, biotechnology, and nanotechnology, for example, witness historical roots going hundreds of years back and new disciplines are not established overnight (Bhushan, 2016; Buchholz & Collins, 2013; Bud, 1994; Ferguson, 2006).

Of the new disciplines mentioned, material science and biotechnology have the oldest roots, and both have ancient Greek origins. Material science can be traced back to metallurgy and ceramics, but the term originated in the 1960s and 1970s. It combines physics, chemistry, and engineering concepts to better understand and develop materials. Materials have become increasingly important in terms of strength, weight, volume, and maintenance as technology has advanced (Ferguson, 2006). The Department of Materials Science and Engineering at the Massachusetts Institute of Technology (MIT) played an important role in formalizing materials science education and research in the 1960s.

Biotechnology can be traced back to fermentation, which has played an important role in the production of food. Later, the biotechnology approach became increasingly important for the production of penicillin and antibiotics, as well as genetic and DNA analysis. As a discipline with journals, education, and so on, it may be difficult to pinpoint the exact date when the first educational program began. MIT had its first graduate program in molecular biology in the 1960s, but biotechnology is a broader concept that was largely established in the 1980s (Buchholz & Collins, 2013).

Hossain et al. (2020) suggests that systems engineering is a relatively new discipline that emerged prior to World War II and grew within the military industry (Hossain et al., 2020). As technology became more complex, the need to view engineering as a system with subsets of areas increased. It was unrelated to fundamental scientific propositions but arose as a result of the growing complexity and need to address embedded societal problems. Today, there are several approaches to system engineering: combining technology and management, requirements and lifecycle, interdisciplinarity, and problem solving (Hossain et al., 2020).

We can argue that all disciplines have evolved into sub-disciplines or subjects in other disciplines. Statistics, a branch of mathematics, has evolved into an auxiliary discipline for other subjects. Chemistry is taught as part of physics in schools, and mathematics is required in all engineering programs because it is one of the fundamental languages of science. Other examples of sub-disciplinary integration will be mathematics, statistics, or physics in engineering.

Digging into the histories of how these disciplines were established reveals a long list of projects, and market and industrial product possibilities. Knowledge integration occurs at various levels, including the establishment of new disciplines across disciplinary boundaries as well as within the boundaries of a discipline (Hacklin & Wallin, 2013).

The development of academic disciplines over time is extremely interesting, with both internal and external market and/or societal drivers. More recently, nanotechnology has emerged as a discipline based on physics and chemistry, with a focus on nanoscales. It can be traced back to the late 1950s with the first educational programs established in the 1990s and early 2000s (Bhushan, 2016). Many younger disciplines, such as artificial intelligence and interaction design, emerged from electronics, computer science, and mathematics (Buchanan, 2005; Cross, 2018; Höök & Löwgren, 2021). All newly established interdisciplinary programs that became disciplines and have academic cross-collaborative historical roots have emerged over time; however, they have done so with a strong request from society and industry.

Newly established programs like techno-anthropology and design-psychology are examples of educational programs across STEM (Science, Technology, Engineering and Mathematics) and SSH (Social Science and Humanities). These educational programs were established in a period where the political system called for new competencies, and one of the criteria for universities to be allowed to start a new program was the market pull and documentation of the new competencies needed.

Experiences from Denmark indicate that the founding of new broader interdisciplinary programs in academia is challenging (Botin et al., 2015). When establishing new broader educations such as this, the experience has shown that candidates face difficulties when entering the labor market. Employers are hesitant to hire candidates with unfamiliar educational backgrounds, even if they have stated in a survey that they require double disciplinary qualifications. Some environmental education programs are also new, but the need for qualified candidates in this field has become much clearer as sustainability issues pervade our society today.

7.4 Engineering and Science

Engineering programs differ in their closeness to science. In a research project on Nordic STEM education, one of the subprojects focused on the development of the future of four existing disciplines: (1) Biotechnology engineering; (2) Mechanical (or industrial economy or production) engineering; (3) Energy engineering; (4) Civil engineering (Routhe et al., 2021). In biotechnology and energy, less change is expected in the future as compared to mechanical and civil engineering. One of the arguments for having a smaller impact on change was its closeness to fundamental science, whereas mechanical and civil engineering had a systemic approach that was determined by society (Gumaelius & Kolmos, 2019). This reminds us that engineering also has numerous facets.

Engineering and science are closely related and share common knowledge constructions that include cause/effect, scale/proportion, quantity research approaches, systems/systems models, structure/function, and stability/change (National Research Council, 2012). Engineering has a scientific foundation, and science cannot develop without technology (Arthur, 2009). However, there are differences that influence knowledge construction and culture.

According to a report from the National Academy of Sciences on STEM education, eight dimensions have different approaches, and it will certainly depend on the type of engineering and science we are considering. Perhaps the differences between biology and biotechnology will be less significant than those between physics and production. Engineering, like science, does not fall into a single category. With inspiration from the eight dimension, described in the National Academy of Sciences, a simplified version of core characteristics are presented in Table 7.1, (National Research Council, 2012).

The questions asked in science and engineering can differ greatly. Scientists will ask questions to discover explanations and causes for specific phenomena, such as what causes cancer or climate change. Engineers will ask questions that are more related to societal problems, with the goal of both understanding the problem and finding the right solutions. Engineers are much more focused on finding solutions

Table 7.1 Comparison of science and engineering

	Science	Engineering
Questions and problems	Starts with a question about a phenomenon	Starts with a problem, need, desire
Explanations and solutions	Explanations	Problem solving
Use of models	Models to develop explanations	Models to analyze systems
Investigations	In the field or in the lab	To gain data or create design criteria
Analyze data	Analyze data for meaning	Analyze data for tests
Mathematics and computational thinking	Fundamental tools	Application for the purpose of design
Evidence	Identifying strength and weaknesses in reasoning	Best solutions
Evaluation and communication	To colleagues	To customer and colleagues

and engaging in creative tinkering, they are fascinated by solution processes. Beauty is involved in both the design and construction.

The scientific method is different. Science creates knowledge in the field and collects data to analyze meaning and causes. Models are tools for systematizing explanations. In engineering, knowledge and data are used to identify design variables, test them, and plan for their implementation. The product is very different because, while scientists are satisfied with theoretical explanations, engineers are concerned with technological development. Of course, there is a mix as engineers require a basic understanding of phenomena before developing designs and solutions. Material science is a discipline that combines the goals of analyzing, understanding, and developing new materials. It is a discipline that has grown from the foundations of physics, chemistry, and mechanical/production.

7.5 The Cultural Dimension of a Discipline

> Disciplines are also cultural phenomena: they are embodied in collections of like-minded people, each with their own codes of conduct, sets of values and distinctive intellectual tasks (Becher, 1981, p. 109).

Another meaning of a discipline is to understand discipline as a culture, and one of the best known books is written by Charles Snow who describes two cultures within the art and the scientific fields (Snow, 1990). As Snow contrasts humanities and science, Kagan builds on a third culture, which is the social sciences as a knowledge domain and practice (Kagan, 2009; Kolmos et al., 2025). As in the previous section, there are many more domains, and engineering might also differ as a culture compared to science.

Becher and Trowler are authors of the famous book *Tribes and Territories,* which is based on empirical data from top universities in the UK and the US. It contends that disciplinary knowledge influences how academics work and research topics, the dissemination patterns, and how personal lives are affected (Becher & Trowler, 2001). The concept of disciplinary practitioners living in tribes with distinct ways of thinking and practicing is part of the academic identity or cultural approach. They emphasize that the disciplines create their own communities through practices that go beyond the professional level, bringing the logic and principles of their work into their personal lives and the formation of identities.

Knowledge is embodied; it is integrated at a personal level and can also be described as a discourse or a community of practice (CoP), which is a group of people who share an interest, experience, value, or concern (Wenger, 1999). It is a group of people who share meaningful practices and have a sense of belonging, creating identities in a domain that gives a sense of belonging (see Chap. 5). The CoP approach can be one that explains a development of disciplines, as the research, teaching practices, and the organization of learning might not automatically be the same in disciplines at different institutions. There are differences in the approaches that in a longer perspective will influence the development of disciplines.

A CoP is an informal organization that has the advantage of being able to bridge various fields and professions by shared interest and that can create social community and discourse. I cannot help but relate to the personal values and life. Pirsig wrote a very charismatic book back in the 1990s called *Zen and the Art of Motorcycle Maintenance*, which is an amazing description of the process of interaction with technology in this case a motorcycle (Pirsig, 1999). One is dragged into a universe of empathy, feelings, values, and love in the interaction with the motorcycle, and it could have been many other technologies or material objects, like a carpenter who cares for the detail in a piece of woodwork or a feeling of creating art.

Maintaining a motorcycle is not quite the same as disciplinary knowledge, but the enthusiasm and the inner description of the emotional relation might very much be the same. It is the empathy, the involvement, the inner joy and pleasure with working processes, writings, constructions, problem solving, etc. For most researchers, writing is a joyful process in which the narrative and the logic slowly fall into place. The review of the literature, finding new literature that contrasts the already embedded, and trying to make it meaningful in an interesting and creative process that in many ways can be compared to painting, where you start with a blank canvas. One color takes over the other and slowly contours emerge, and maybe it takes a different direction than in the beginning. It could also be the lab experiment that produces surprising results. The writing, painting and the lab experiment can all be situations where you become one with your action for a while.

7.6 Approach to Learning

According to Barnett, STEM and humanities curricula vary significantly. He compares the three overall elements of the curriculum: knowledge, action, and the self, noting that knowledge may be equally important in STEM and humanities, but action and the self differ significantly. Action is important in STEM and there is less emphasis on the self and identity development, whereas in the humanities, the self is more important than the actions (Barnett & Coate, 2004). Actions can be translated into skills, and the self leads to personal reflection and identity development. This is embedded in the culture, the learning, the specific curriculum, and the discourses. If the self is never articulated as an important component in reflection and learning, it remains an inner and potentially un-reflected process.

In a review on teaching generic competencies, we found that the learning of problem identification, project management (including the understanding of project phases and sharing of work), collaboration, and communication are all saturated by the discipline (Kolmos et al., 2025). This will impact interdisciplinary collaboration if the students do not share the knowledge and ways to collaborate, which further complicates the interdisciplinary collaboration. The question is how generic the generic competencies are, a question that is raised in several studies (Chan & Fong, 2018; Jones, 2009, 2013).

If social science has an emphasis on problem identification as the main objective, and engineering focuses on the problem solutions, there is a mishmash of the priorities and expectations for collaboration. Differences and expectations need to be articulated. Tacit knowledge is invaluable in learning, as it is part of intuition and creativity. However, in an interdisciplinary collaborative situation, we have to find some shortcuts to enhance the collaborative learning process to gain common understandings.

The disciplinary differences in academia also match professional differences, which makes the learning of interdisciplinary collaboration even more important in higher education (Clarke et al., 2013; Gordon & Whitchurch, 2010). In the literature on employability, one of the voices from industry outlines the generic competency portion that the young engineers can learn in their first job; another mentions that the generic competency part is what matters, and the specific knowledge can come later. I would argue that from a learning point of view, the practice competencies are not learned by experiencing only one situation, as discussed in Chap. 6. Students must experience this several times, and studies clearly indicate that interdisciplinary collaboration is fundamental in engineering practice (Holgaard et al., 2025).

Approaches to learning differ according to disciplinary and to cultural practices. An example of cultural practices could be when electronics engineering started using PBL in the 1990s. One institution argued that doing projects was the only way to teach and learn electronics, while another argued that you could only learn electronics through lectures and regular classroom teaching with smaller individual tasks. Both the teaching and learning will be influenced by the disciplinary and institutional cultures.

I made this detour to get a better grasp on what it is we must learn to cross, combine, or integrate? As disciplines are not only knowledge but learning and culture as well, it is important to understand all elements in the process in interdisciplinary collaboration. Participants in interdisciplinary projects should be aware of knowledge, the understanding of learning and generic competencies embedded in one's own discipline, as well as cultural differences (see Fig. 7.1).

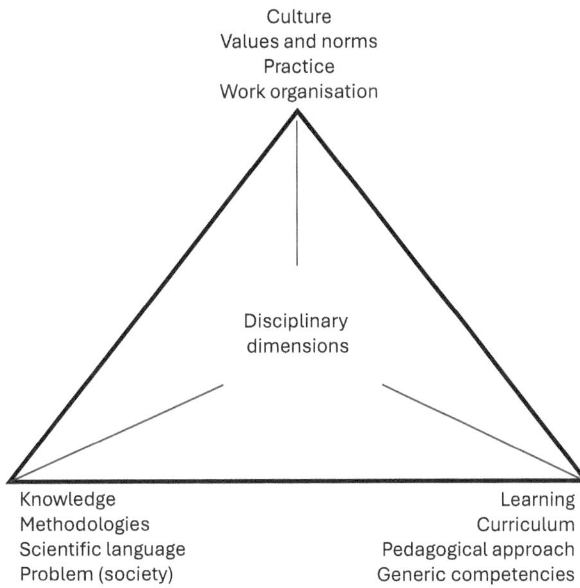

Fig. 7.1 Dimensions of interdisciplinary collaboration (Kolmos, Holgaard and Routhe, 2025)

We need to have this in mind when we facilitate students' interdisciplinary learning processes, and they need to address it in their approach to collaboration. This more structured understanding of interdisciplinary collaboration, based on the variables in disciplines, can further develop interdisciplinary competencies and learnings or at least a comprehensive understanding of the issues (see Chap. 8).

7.7 What is Interdisciplinarity?

If I turn to the understanding of interdisciplinarity, there are again several definitions. An often-applied understanding points at the following characteristics:

> Interdisciplinarity tackles questions (problems, themes) that are (or at least might be) addressed by multiple disciplines (or 'knowledge communities' more generally).
> Interdisciplinarity seeks to integrate the insights from multiple disciplines in order to generate a superior understanding of a particular question.
> Interdisciplinarity is thus open to theories, methods, philosophical perspectives, and types of data utilized in different disciplines (Szostak et al., 2016, p. 1).

The concept of interdisciplinarity indicates that disciplines are 'stable' constructions and there is some form of interaction between two or more disciplines (Hvidtfeldt, 2018). What kind of interaction? That is a good question! There is an agreement that there are wide definitions and concepts such as the most often used multidisciplinarity, interdisciplinarity, or transdisciplinarity, and several other concepts.

Julie Klein is one of the most cited in this area and she indicates that interdisciplinarity is:

> neither a subject matter nor a body of content. It is a process for achieving an integrative synthesis, a process that usually begins with a problem, question, topic, or issue (Klein, 1990, p. 189)

Multidisciplinarity is the approach where disciplines are placed side by side and you can discuss borrowing knowledge from each other's areas. Hacklin and Wallin make a distinction between disciplinary knowledge integration within discipline boundaries and interdisciplinary knowledge integration across discipline borders, which might be within a multidisciplinary area where the disciplines might not have equal status (Hacklin & Wallin, 2013). The issue is that many practices are called interdisciplinary without any critical reflection if there really is an *integrative* process. Klein emphasizes that there may be constant boundary crossings or cross fertilization between disciplines that borrow tools, methods, or concepts from each other; however, this is not necessarily interdisciplinarity (Klein, 2006).

In Klein's understanding, interdisciplinarity is an integrative process that can be synonymous with synthesis (Szostak et al., 2016). Klein makes distinctions between four levels of integration. Contextual information is shared in the first level, while discoveries are explained in the second. Rather than being really integrated, both levels are more multidisciplinary. Deeper conceptual understanding is shared in the next two levels, and as for the 'real integrative level', questions are redefined by the participating disciplines (Klein, 2010; Szostak et al., 2016). An integration process will always be fluent where there will be periods of closer collaboration and periods of more distant collaboration.

Keestra and Menken have illustrated the integration process by two parallel lines to illustrate a working process, where several points cross each other or merge for periods, which is a real collaborative process (Keestra & Menken, 2016). In an interdisciplinary project, the scope of the problem calls to involve more disciplines and thereby adds to a common goal. The process for periods can be parallel, like in multidisciplinary collaboration or more integrated in other periods. However, it is not the type of collaboration determining if it is multi- or interdisciplinary; it is the problem and the goal that calls for the involved disciplines. The process surely affects the product and outcome, but it is a means to achieve a purpose.

What might be important are the disciplines that the problem calls for. Klein has made a very valuable distinction between narrow and broad interdisciplinarity (Klein, 2010). The narrow interdisciplinary approach consists of disciplines within similar knowledge domains, such as biotechnology, chemistry, biology, and possibly chemical engineering. Within the engineering disciplines there are plenty of disciplinary combinations in a technological development, which are also shared with science disciplines. These share knowledge paradigms in the world view, theoretical foundations and methodologies. Students working across these boundaries do have some common ground, although the pedagogy and the culture might be different.

The broad interdisciplinary approach is much more difficult as it will have to cross the cognitive knowledge boundaries as well as cultural and pedagogic differences. Communication and understanding of different discourses become even more important in the collaborative interaction (Choi & Richards, 2017).

Transdisciplinarity has different definitions. One definition is that knowledge, ideas, methods and products go beyond the existing disciplines in opening up new horizons or new disciplines. This will involve a re-structuring of the knowledge creation process and open new horizons (Klein, 2006; Pennington, 2016). The most widespread understanding today is that collaborations somehow transcend academia and that there is some kind of collaboration with external partners (Keestra & Menken, 2016; Pennington, 2016). In my use of transdisciplinarity, it is the collaboration between academia and non-academia. We cannot expect engineering students to be the drivers for creating new integrated interdisciplinary knowledge which involve an emerge of now knowledge constructions. This is a task for academics.

When comparing the understandings of interdisciplinarity with what I found in my search for what a discipline can be characterized as, I found many differences. The cultural and pedagogical angles are not explicitly present in the notions of interdisciplinarity and even if interdisciplinarity should not be defined by collaboration, this dimension will influence the integration processes. In education, this will be an important aspect as it is about learning to identify the boundaries and finding ways to cross the boundaries.

7.8 Interdisciplinary Projects

The title of this chapter is 'Interdisciplinary Collaboration Within or Across Programs'. I think that I have already given my answer to this question, as new interdisciplinary programs could become disciplines at a certain time, like the establishment of many of the 'younger' engineering programs. Broader interdisciplinary programs combining engineering, humanities, or social science are needed from a societal and human perspective; however, the candidates do have difficulties in getting jobs, at least in Denmark.

In the early PBL literature from the Danish reform universities (see Chap. 3) and in the characteristics of project types, there has been interdisciplinary project that has been defined (Kolmos, 1996). During the last years at conferences, I became concerned that many engineering institutions claimed to run interdisciplinary projects, which indeed sometimes only included software, computer science, and electronics, for example. Authors and presenters claimed that this was really a challenging process to get these programs to work together, and maybe indeed this has been the case. Did institutional frames, culture, and learning approaches actually cause the challenges and not so much the cognitive knowledge perspectives?

It was one of the intentions of the research project *Interdisciplinary PBL methodologies in engineering education and work* (InterPBL) to analyze and categorize interdisciplinary projects. Based on action research, we interacted and followed the development of interdisciplinary projects at Aalborg University and created a model

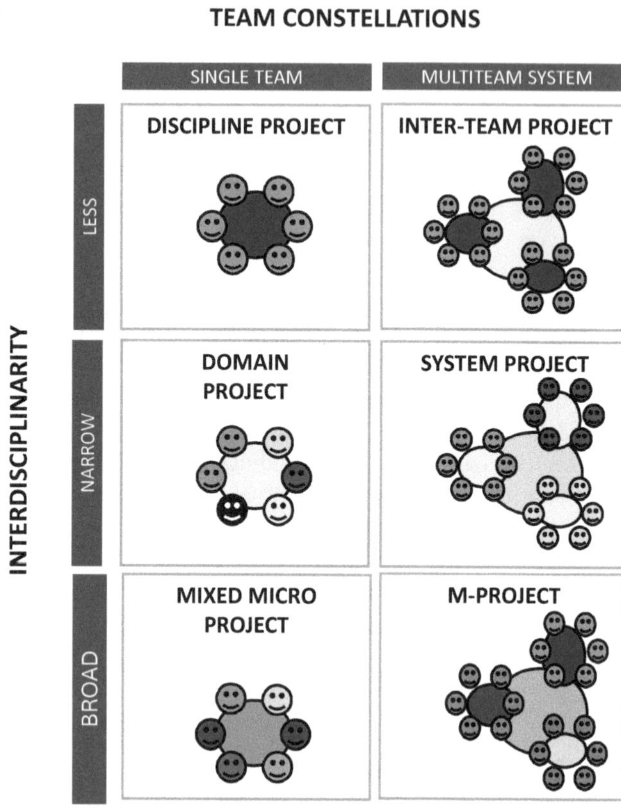

Fig. 7.2 Interdisciplinary project types combined with collaboration patterns (Kolmos et al., 2025)

of them (Fig. 7.2). We have two dimensions, the knowledge with progression from disciplines to narrow and broad interdisciplinary approaches, and team dimension with single teams or teams in network or multiteam systems (Kolmos et al., 2024).

The concept of interdisciplinarity can be both narrow and broad. The choice of disciplines will depend on the purpose of the problem. The narrow interdisciplinary approach most often leads to construction of technological systems. The broad interdisciplinary approach, which spans the humanities, social sciences, engineering, science, medicine, and health, is the other dimension. That is a challenge, as there are differences in worldviews, theoretical foundations, theories, methodologies, etc. To approach problems and projects, students in a broad interdisciplinary project team must first teach one another about their own discipline and worldview.

In philosophy, the main purpose might be to understand the problem, whereas engineering students prefer to jump to solutions, preferably immediately. In the collaboration, there will be differences in knowledge, culture, and pedagogy. There will be pedagogical, cultural, and knowledge gaps in the partnership. The discourses and topics discussed are shaped by the knowledge paradigms. In contrast to engineering, which has a much more evident focus on the product, humanity is frequently far more process- and individual growth-minded, so the focus on the product and process will be different.

The team dimension covers single teams versus teams of teams. Nearly all of the implementation of PBL is in single teams, typically in one course in a classroom setting (Chen et al., 2023). Operating with teams of teams, or what is also known as multiteam systems or inter-team collaboration, is relatively new for education. In the engineering profession, it is very common in software development, product development, or other technological innovations (Nielsen et al., 2010).

There is significant literature on, for example, megaprojects working on complex problems. The characteristics of megaprojects can address complex problems, interdisciplinary (both broad and narrow) large investments, and large organizations to implement the project. These projects can be very hard to control as there are multiple teams working together (Hu et al., 2015). Megaprojects are mostly longitudinal with also great collaborative complexity. They can be large infrastructure projects in cities, high-speed trains, large airports, development of new aircrafts, space technologies, etc. (Brown, 1993; Priemus et al., 2008).

Combining these two dimensions gives six project types, a kind of taxonomy. It gives a language to speak with more variation about projects in engineering education and the involved collaboration patterns. We have identified discipline projects both in single teams and in inter-team constructions, especially in software development and computer science (Graham, 2022).

Narrow interdisciplinary projects are the most common types in engineering education both for single teams and multiteam compositions, and the context for a lot of the research literature on interdisciplinary learning (Beddoes & Borrego, 2014; Lattuca et al., 2013; Lattuca et al., 2017). The multiteam category, system projects, is also normally used for software or technology development. In the InterPBL project we have researched such project types by identifying learning outcomes that go far beyond the formal curriculum (Routhe et al., 2023; Routhe et al., 2024). These practiced system projects consist of disciplinary project teams working across the disciplines to construct new technologies. The findings from that research indicate that there is a move from learning project management competencies in single teams to learning both management and coordination/leadership competencies for the inter-team projects.

The last types of projects are within the broader interdisciplinary perspective, and we have called these projects mixed micro-projects for single teams and M-projects for the multiteam constructions. There are practices for both categories in the literature; however, the reason for calling the first category ‘mixed micro’ is that it seems to be team practices within a limited time schedule, which could make sense. The broad focus in a single team will have limited resources, and the learning outcome might not be to create a product but to gain an idea of other disciplines (Routhe et al., 2022). M-projects can mean mega-projects or mission projects and are broad, interdisciplinary, and multiteam. Aalborg University has managed to create interdisciplinary mega-projects, however with limited success (Bertel et al., 2021; Kolmos et al., 2024).

Our intention was to make a categorization of interdisciplinary projects. As soon as we create a language, we establish boundaries and limits to existing practice. Perhaps a project was intended to be a system project, but through the students’ learning process it evolved into a multiteam project with a bigger disciplinary approach. It could also be the other way around: a system project suddenly expands, and students begin collaborating with fields in humanities or social science. This categorization, therefore, is the starting point of a student's journey.

This categorization does not include transdisciplinarity. For students' projects it could by adding a third column that is about collaborating with students from non-academic educations or with non-academic background. The transdisciplinary perspective will be important in learning to respond to climate chaos as craft might be crucial in such situations.

During the last 10 years, we also see more literature reporting on multi-, inter- and transdisciplinary capstone projects (Zheng et al., 2021). Real life or authentic problems normally need knowledge across the disciplines in both the problem analysis and problem-solving phases. The categorization of interdisciplinary projects is not intended as a progression in the sense that students need to go through the disciplinary and narrow interdisciplinary projects to start the more complex broad interdisciplinary projects, although research indicates that students would like to have some kind of scaffolding and guidance in analyzing and solving complex problems as they often find it difficult (Berg et al., 2025; Winther et al., 2022).

Even if the students have learned collaboration and management in a disciplinary project, the learning of generic competencies is contextual, and students might have difficulties in transforming their learning from one context to another different one (Kolmos et al., 2024). Thus, there needs to be a facilitation of the student learning process, no matter if it is in disciplinary or broad interdisciplinary projects. I will discuss this in much more in detail in Chap. 8.

7.9 Interdisciplinary Learning Outcomes and Collaboration

The categorization of interdisciplinary projects has gained attention, and the intention is to become more aware of conceptualizations of interdisciplinary projects and indeed of a greater variation of learning outcomes, which is a core element in educational design.

For learning outcomes, it can be summed up as this: the generic competencies are means in learning the disciplines in the disciplinary projects; however, for the interdisciplinary projects, the disciplinary knowledge is a means to learning interdisciplinary collaborative competencies (see Fig. 7.3) (Kolmos et al., 2025).

This understanding of learning outcomes in disciplinary and interdisciplinary projects adds to a more detailed approach. There is also increased emphasis on the components of collaboration as shown in Fig. 7.3. The need for mutual accountability, ownership, collective work products, trust in all aspects, the creation of shared mental modelling, and the ability to solve social and professional conflicts will all increase from disciplinary to broad interdisciplinary projects. As the problems and tasks become more comprehensive, the teams become larger and there is

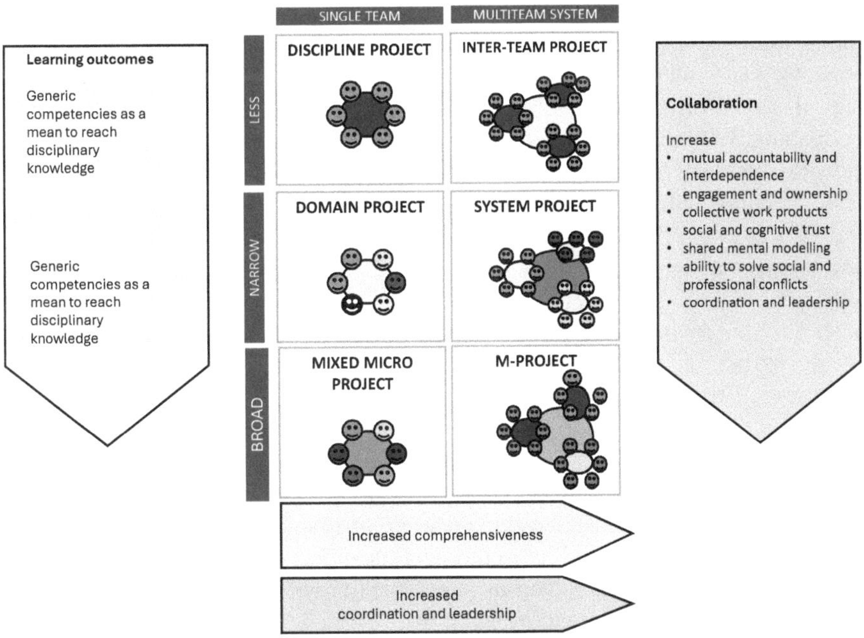

Fig. 7.3 Project types, learning outcomes and collaboration

an increased need for coordination and leadership as competencies among the students (Routhe et al., 2024).

Lattuca and Knight (2010) define the interdisciplinary competencies with a focus on knowledge and professional skills. Awareness of disciplinary knowledge, appreciation of disciplinary perspectives, and recognition of disciplinary limitations are some of the core interdisciplinary competencies for collaboration together with a series of competencies to collaborate (Lattuca et al., 2017). As mentioned in the previous sections, the boundaries for a discipline can be very blurred, and it can be hard to distinguish between programs such as industrial production and mechanical engineering, or mechanical engineering and material science. Each discipline will have an overlap with the other and the boundaries can be fuzzy. In the findings from the InterPBL project and the analyses of students' experience of system projects, students indicate that they learn more about their own discipline and the boundaries by identifying what their discipline is not (Routhe et al., 2023).

Disciplinary grounding is another concept used in the literature to indicate that you can only be 'inter' if you have a solid starting point (Tripp & Shortlidge, 2019). Tripp and Shortlidge emphasize this point as a more basic but necessary approach to create interdisciplinary collaboration, namely disciplinary humility. This is along the same lines as Richter and Paretti who are working with the concept of disciplinary egocentrism, which has to be avoided in interdisciplinary collaboration and is the opposite of disciplinary humility (Richter & Paretti, 2009).

Transitioning from disciplinary to interdisciplinary projects undoubtedly affects collaboration, and work needs to be done to develop shared mental models of the project's goals and direction as well as the interactions that occur during the process. Based on original research and a review of the literature, Pennington created a collaborative model that considers the system, the collaborative process, and the individual learner. She recognizes several critical variables in the integration process, including understanding each other's viewpoints, engagement and ownership, and flexible problem identification (Pennington, 2016). This is consistent with the development of interdisciplinary competencies for collaboration, which include reflection, the ability to identify individual and team biases, the appreciation of other disciplinary perspectives, and a willingness to find common ground (Lattuca et al., 2013), as well as the concept of disciplinary humility (Tripp & Shortlidge, 2019).

The learning outcomes are one of the reasons why academic staff are not necessarily in favor of interdisciplinary projects, as the focus for learning will change from disciplinary deep knowledge to interdisciplinary generic competencies. However, arguments against the statement on learning outcomes are that students learn about their own discipline in the interdisciplinary collaboration. Academic staff who are positively involved in interdisciplinary learning also have to cross the boundaries and experience how it feels which can be as hard (Järvinen et al., 2025).

To design interdisciplinary learning, the understanding of a discipline emphasizes knowledge, culture and learning, and these variables will influence the interdisciplinary collaboration and include questions such as (Kolmos et al., 2025, p. 11):

- Which disciplines do we bring together?
- What knowledge and epistemologies do they bring?
- What is the relationship to society and the understanding of problems?
- What is the reflection on the culture, societal values, and work values?
- How do we understand learning from experiential to taught textbook knowledge?
- How are the projects organized? How much does the problem count in the projects?
- How is the work organized?
- How do the sharing, collaboration, structure, and coordination happen in the project?

Figure 7.4 illustrates how a broader approach is required for designing, planning, and facilitating interdisciplinary learning. In addition to the disciplines, there will be relationships between various cultural and pedagogical components.

Every element, including knowledge, culture, and learning, will influence the dimensions of the other discipline. The fact that we not only deal with three dimensions will exacerbate the chaos. Culture includes values and norms, as well as work organization and learning, which go beyond pedagogy and generic competencies. Knowledge encompasses theories, methodologies, and societal relationships.

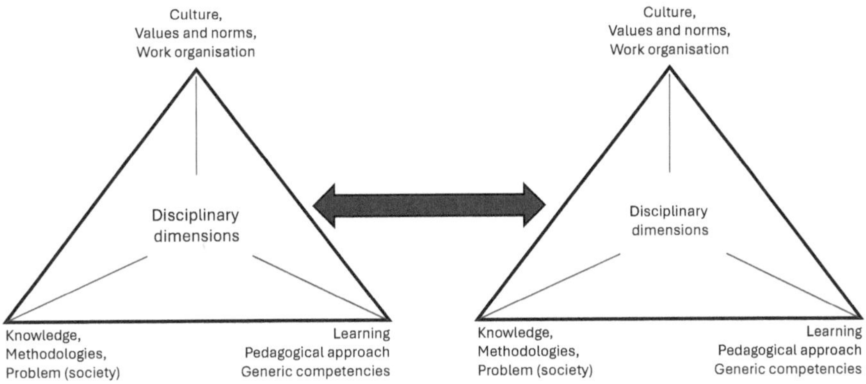

Fig. 7.4 Relationships among the disciplinary dimensions for two disciplines (Kolmos et al., 2025)

7.10 Reflection 7: Living-with Chaos

Interdisciplinary collaboration is essential in learning to navigate a more chaotic world. There is a need to cross disciplinary boundaries, regardless of how we define and approach disciplines. Disciplinary boundaries exist, whether they are understood as epistemological knowledge constructions or as institutional and cultural components, and we must find ways to link them.

As learners, we must be adaptable, and the ideal situation is one in which we are aware of our own knowledge contributions while remaining open to learning from others. In an educational setting, these are high expectations because the learner is forming their identity and gaining confidence in knowledge, competencies, and attitudes. On the other hand, education provides us with the opportunity to teach new types of boundaries that not only frame existing disciplines but integrate them with interactions with other disciplines. If this is the case, and the academic staff are willing to facilitate these processes from both a knowledge and learning standpoint, I am confident that we can educate humane capacities for the future. This is in the meaning of not only what the society needs, it is also what the individual needs for learning to live-with chaos.

At a conference following a keynote on interdisciplinary projects, I was asked if we could imagine a society without boundaries. My response was that I didn't think we could. We require boundaries in order to form our identities and sense of community belonging. A society without boundaries will necessitate very strong individuals who understand their worth. We need to interact with the academic society and feel like we belong to a community. The question is whether we will stick to disciplinary institutionalized boundaries or try to create crossover tracks for students.

This may be simple to implement in elective curriculum systems, as there are options for offering compulsory interdisciplinary courses with the learning outcome of gaining experience with interdisciplinary projects and learning the competencies required to work in these settings. This is a domain project made up of individual students from various disciplinary backgrounds, and each student must be accountable for his or her disciplinary knowledge while also participating in an interactive interdisciplinary process. In institutionalized PBL curricula, as described in Chaps. 3 and 4, PBL is organized into disciplines and silos as the primary approach, and this might be more complicated.

As demonstrated in this chapter, it is possible to design interdisciplinary projects such as domain projects, system projects, M-projects, and mixed-micro projects. This is a new task for universities that goes beyond traditional learning outcomes and involves the development of resilience and well-being in the context of agency and mastery. More on this in Chap. 8.

References

Abbott, A. (2002). The disciplines and the future. In *The future of the city of intellect: The changing American university* (pp. 205–230).

Arthur, W. B. (2009). *The nature of technology: What it is and how it evolves*. Simon and Schuster.

Barnett, R., & Coate, K. (2004). *Engaging the curriculum in higher education* (1 edition ed.). Open University Press.

Becher, T. (1981). Towards a definition of disciplinary cultures. *Studies in Higher Education, 6*(2), 109–122.

Becher, T., & Trowler, P. (2001). *Academic tribes and territories: Intellectual enquiry and the cultures of disciplines* (2nd ed.). Open University Press.

Beddoes, K., & Borrego, M. (2014). Facilitating formation of shared mental models in interdisciplinary graduate student teams. *International Journal of Collaborative Engineering, 1*(3–4), 236–255.

Berg, V., Wallin, P., Sjursen, R. L., & Bolstad, T. (2025). New approach and old challenges: Challenging engineering students' interdisciplinary understanding through interdisciplinary projects and resources for disciplinary student groups. *European Journal of Engineering Education*, 1–19.

Bertel, L. B., Winther, M., Routhe, H. W., & Kolmos, A. (2021). Framing and facilitating complex problem-solving competences in interdisciplinary megaprojects: An institutional strategy to educate for sustainable development. *International Journal of Sustainability in Higher Education., 23*, 1173–1191.

Bhushan, B. (2016). Introduction to nanotechnology: History, status, and importance of nanoscience and nanotechnology education. In *Global perspectives of nanoscience and engineering education* (pp. 1–31).

Botin, L., Bertelsen, P., & Nøhr, C. (2015). *Techno-anthropology in health informatics: Methodologies for improving human-technology relations* (Vol. 215). IOS Press.

Brown, R. B. (1993). Meta-competence: A recipe for reframing the competence debate. *Personnel Review., 22*, 25–36.

Buchanan, B. G. (2005). A (very) brief history of artificial intelligence. *AI Magazine, 26*(4), 53–53.

Buchholz, K., & Collins, J. (2013). The roots—a short history of industrial microbiology and biotechnology. *Applied Microbiology and Biotechnology, 97*, 3747–3762.

Bud, R. (1994). *The uses of life: A history of biotechnology*. Cambridge University Press.

Chan, C. K., & Fong, E. T. (2018). Disciplinary differences and implications for the development of generic skills: A study of engineering and business students' perceptions of generic skills. *European Journal of Engineering Education, 43*(6), 927–949.

Chen, J., Du, X., Chaaban, Y., Velmurugan, G., Lyngdorf, N. E. R., Nørgaard, B., Routhe, H. W., Hansen, S., Guerra, A., & Bertel, L. B. (2023). An exploration of sources fostering first-year engineering students' academic well-being in a PBL environment. *IEEE Transactions on Education, 66*(5), 421–430.

Chettiparamb, A. (2007). *Interdisciplinarity: A literature review*. University of Southampton.

Choi, S., & Richards, K. (2017). *Interdisciplinary discourse: Communicating across disciplines*. Springer.

Clarke, M., Hyde, A., & Drennan, J. (2013). Professional identity in higher education. In *The academic profession in Europe: New tasks and new challenges* (pp. 7–21).

Cross, N. (2018). Developing design as a discipline. *Journal of Engineering Design, 29*(12), 691–708.

Fensham, P. J. (2004). *Defining an identity: The evolution of science education as a field of research* (Vol. 20). Springer Science & Business Media.

Ferguson, C. (2006). Historical introduction to the development of material science and engineering as a teaching discipline. *Available from the UKCME at http://www.materials.ac. UK/pub/ Materials-History-Intro. pdf.*

Gordon, G., & Whitchurch, C. (2010). *Academic and professional identities in higher education*. Routledge.

Graham, R. (2022). *CEEDA case study report: Six case studies of the impact of COVID-19 on global practice in engineering education.*

Gumaelius, L., & Kolmos, A. (2019). The future of engineering education: Where are we heading. In *SEFI 47th Annual Conference: Varietas Delectat, Proceedings of the Complexity is the New Normality*. Hungary.

Habbal, F., Kolmos, A., Hadgraft, R. G., Holgaard, J. E., & Reda, K. (2024). *Reshaping Engineering Education: Addressing Complex Human Challenges*. Springer Nature.

Hacklin, F., & Wallin, M. W. (2013). Convergence and interdisciplinarity in innovation management: A review, critique, and future directions. *The Service Industries Journal, 33*(7–8), 774–788.

Holgaard, J. E., Kolmos, A., Friedrichsen, D., & Routhe, H. W. (2025). Interdisciplinarity, generic competencies, and traits in engineering practice from the perspective of early-career engineers. *European Journal of Engineering Education*, 1–17.

Hossain, N. U. I., Jaradat, R. M., Hamilton, M. A., Keating, C. B., & Goerger, S. R. (2020). A historical perspective on development of systems engineering discipline: A review and analysis. *Journal of Systems Science and Systems Engineering, 29*, 1–35.

Hu, Y., Chan, A. P., Le, Y., & Jin, R.-z. (2015). From construction megaproject management to complex project management: Bibliographic analysis. *Journal of Management in Engineering, 31*(4), 04014052.

Hvidtfeldt, R. (2018). *The structure of interdisciplinary science*. Springer.

Höök, K., & Löwgren, J. (2021). Characterizing interaction design by its ideals: A discipline in transition. *She Ji: The Journal of Design, Economics, and Innovation, 7*(1), 24–40.

Jones, A. (2009). Redisciplining generic attributes: The disciplinary context in focus. *Studies in Higher Education, 34*(1), 85–100.

Jones, A. (2013). There is nothing generic about graduate attributes: Unpacking the scope of context. *Journal of Further and Higher Education, 37*(5), 591–605. https://doi.org/10.1080/0309877X.2011.645466

Järvinen, M., Virtaluoto, J., Lehesvuori, S., Roslöf, J., & Hämäläinen, R. (2025). Faculty experiences of boundary crossing in an interdisciplinary degree program. *European Journal of Engineering Education*, 1–19.

Kagan, J. (2009). *The three cultures: Natural sciences, social sciences, and the humanities in the 21st century*. Cambridge University Press.

Keestra, M., & Menken, S. (2016). *An introduction to interdisciplinary research: Theory and practice*. Amsterdam University Press.

Klein, J. T. (1990). *Interdisciplinarity: History, theory, and practice*. Wayne State University Press.

Klein, J. T. (2006). *A platform for a shared discourse of interdisciplinary education*.
Klein, J. T. (2010). A taxonomy of interdisciplinarity. In *The Oxford handbook of interdisciplinarity* (Vol. 15, pp. 15–30).
Kolmos, A. (1996). Reflections on project work and problem-based learning. *European Journal of Engineering Education, 21*(2), 141–148. https://doi.org/10.1080/03043799608923397
Kolmos, A., Holgaard, J. E., & Routhe, H. W. (2025). Understanding and designing variation in interdisciplinary problem-based projects in engineering education. *Education Sciences, 15*(2), 138.
Kolmos, A., Holgaard, J. E., Routhe, H. W., Winther, M., & Bertel, L. (2024). Interdisciplinary project types in engineering education. *European Journal of Engineering Education*, 1–26.
Lattuca, L. R., Knight, D., & Bergom, I. (2013). Developing a measure of interdisciplinary competence. *International Journal of Engineering Education, 29*(3), 726–739.
Lattuca, L. R., Knight, D. B., Ro, H. K., & Novoselich, B. J. (2017). Supporting the Development of Engineers' Interdisciplinary Competence. *Journal of Engineering Education., 106*, 71–97.
Millar, V. (2019). Interdisciplinary curriculum reform in the changing university. In *Curriculum as Contestation* (pp. 111–123). Routledge.
Ming, X., MacLeod, M., Boon, M., & van der Veen, J. (2025). Envisioning and interpreting interdisciplinary engineering education (IEE): Different dimensions of understanding interdisciplinarity. *European Journal of Engineering Education*, 1–21.
National Research Council. (2012). *A framework for K-12 science education: Practices, crosscutting concepts, and core ideas*. National Academies Press.
Nielsen, J. F. D., Du, X., Kolmos, A., Nielsen, J. F. D., Du, X., & Kolmos, A. (2010). Innovative application of a new PBL model to interdisciplinary and intercultural projects. *International Journal of Electrical Engineering & Education, 47*(2), 174–188.
Pennington, D. (2016). A conceptual model for knowledge integration in interdisciplinary teams: Orchestrating individual learning and group processes. *Journal of Environmental Studies and Sciences, 6*(2), 300–312.
Pirsig, R. M. (1999). *Zen and the art of motorcycle maintenance: An inquiry into values*. Random House.
Priemus, H., Flyvbjerg, B., & van Wee, B. (2008). *Decision-making on mega-projects: Cost-benefit analysis, planning and innovation*. Edward Elgar Publishing.
Richter, D. M., & Paretti, M. C. (2009). Identifying barriers to and outcomes of interdisciplinarity in the engineering classroom. *European Journal of Engineering Education, 34*(1), 29–45.
Routhe, H. W., Holgaard, J. E., & Kolmos, A. (2023). Experienced learning outcomes for interdisciplinary projects in engineering education. *IEEE Transactions on Education*, 1–13. https://doi.org/10.1109/TE.2023.3284835
Routhe, H. W., Holgaard, J. E., & Kolmos, A. (2024). Students' learning of management and leadership in engineering education–a literature review. *European Journal of Engineering Education, 49*(3), 540–576.
Routhe, H. W., Winther, M., Magnell, M., Gumaelius, L., & Kolmos, A. (2021). Faculty perspectives on Future Engineering Education. In *Proceedings of REES AAEE 2021*. The University of Western Australia, Perth.
Routhe, H. W., Winther, M., Nguyen, N. T., Holgaard, J. E., & Kolmos, A. (2022). *Challenges for engineering students working with authentic complex problems*. SEFI 2022: Annual Conference 19–22 September 2022.
Ryberg, T., & Kolmos, A. (2023). Hybridisation, variation and reflection: New PBL curriculum principles. In *PBL in a Digital Age* (pp. 27–38). Aalborg Universitetsforlag.
Snow, C. P. (1990). The two cultures. *Leonardo, 23*(2/3), 169–173.
Szostak, R., Gnoli, C., & López-Huertas, M. (2016). *Interdisciplinary knowledge organization*. Springer.
Tripp, B., & Shortlidge, E. E. (2019). A framework to guide undergraduate education in interdisciplinary science. *CBE Life Sciences Education, 18*(2), es3. https://doi.org/10.1187/cbe.18-11-0226

Wenger, E. (1999). *Communities of practice: Learning, meaning, and identity*. Cambridge University Press.
Winther, M., Routhe, H. W., Holgaard, J. E., & Kolmos, A. (2022). *Interdisciplinary problem-based projects for first-year engineering students*. ASEE 2022 Anual Conference: Excellence through Dicersity.
Yates, L., Woelert, P., Millar, V., & O'Connor, K. (2016). *Knowledge at the crossroads?: Physics and history in the changing world of schools and universities*. Springer.
Zheng, L., Hu, D., & Jesiek, B. K. (2021). *A systematic review of multidisciplinary engineering education: Accredited programs, educational approaches, and capstone design*. 2021 ASEE Virtual Annual Conference Content Access,

Chapter 8
Boundary Crossing, Transformation, and Emotions

8.1 Marketplace for Interdisciplinary Competencies

I mentioned in Chap. 6 that there is a tendency for the lists of competencies to grow and develop. When it comes to interdisciplinary abilities, these lists appear to grow even longer after only reading a portion of the literature. The language varies between studies and the meaning can overlap. Still, it is a requirement for any program designer to identify the essential parts in interdisciplinary competencies.

First and foremost, there are new broad notions on the table. Repko et al. (2019) distinguished between mentality, attributes, and competencies. Empathy, ethics, humility, appreciation for variety, and civic participation are all examples of mindset elements (Repko et al., 2019). The skills highlighted include communication, abstract and creative thinking, and metacognition. Traits such as a passion for learning, self-reflection, intellectual daring, and a willingness to take risks are all examples of important characteristics.

All these characteristics, talents, or mindsets may be difficult to include into a curriculum because they often pertain to a personal rather than educational level. What can be defined as learning objectives in education should be examined, and assessing love of learning and humility might be very challenging. Reflection, both collaborative and individual, can undoubtedly be included in a curriculum, but it is difficult to separate in a grading process.

A recent study on young engineers' experiences with interdisciplinarity in their professional practice found that personal traits are essential in what respondents defined as interdisciplinary work (Holgaard et al., 2025). Personal attitudes are vital in an interdisciplinary setting because one will encounter novel conditions, scientific terminology, cultural habits, and learning methodologies. It is like meeting an entirely different culture and learning to interpret and react. I am still surprised that it took me so long to realize that, for example, in a Muslim culture you do not often

A. Kolmos, *Engineering Education for Interdisciplinary Futures*, Emergent Directions in Engineering Education 2,
https://doi.org/10.1007/978-3-032-32540-2_8

shake hands across genders and that I had to wait to see how I should greet in new settings.

Tripp and Shortlidge identify four elements important for interdisciplinary understanding: disciplinary grounding, research methods, knowledge integration, and collaboration. The process to gain interdisciplinary understanding is disciplinary humility, which is a personal quality to be open to other fields and what Repko calls 'mindset' (Tripp & Shortlidge, 2019). Lattuca et al. (2013) is a highly cited reference identifying eight competencies that can be grouped in Tripp and Shortlidge's framework (see Table 8.1).

Interdisciplinary communication, trust, active listening, and risk-taking are also elements that can facilitate or hinder the collaborative aspect (Beddoes, 2020; Borrego & Cutler, 2010). Competencies or traits such as humility, shared knowledge, shared mental models, and interdisciplinary collaboration are all aspects of interaction and integration of knowledge, methodologies, and epistemologies. This is echoed in the study by Claus and Wiese (2019) who, at an individual level, identify four overall dimensions of interdisciplinary competencies that characterize the extra attention given to collaboration in an interdisciplinary setting: initiative for exchange, target group-specific communication, knowledge integration, and reflection and appreciation (Claus & Wiese, 2019).

Bronstein's model for interdisciplinary collaboration includes other dimensions: new professional activities, collective ownership, interdependence, flexibility, and reflection (Laura R. Bronstein, 2003). Bronstein emphasizes that interdisciplinary collaboration is only needed when there is a new issue that needs to be analyzed or solved, and therefore the new professional activity is a foundation

Table 8.1 Comparing Tripp and Shortlidge's framework with Lattuca et al.'s eight competencies

Tripp and Shortlidge (Tripp & Shortlidge, 2019)	Lattuca et al. (Lattuca et al., 2013)
Disciplinary grounding and different research methods	Awareness of disciplinary knowledge, such as being aware of disciplinary boundaries and approaches.
	Appreciation of disciplinary perspectives, including an appreciative attitude towards other stories and frames of reference.
	Recognition of disciplinary limitations, which includes critical thinking skills.
Advancement through integration	Interdisciplinary evaluation in order to increase the quality of interdisciplinary work.
	Reflexivity, including the ability to reflect on one's biases and choices when handling complex problems.
Collaboration across disciplines	The ability to find common ground as a basis for collaboration work is crucial.
	Integrative skill and the capacity to integrate knowledge and modes of thinking.
Disciplinary humility	Appreciation of non-disciplinary perspectives, which relates to appreciation of transdisciplinary relations.

for interdisciplinary collaboration together with the feeling of having shared ownership and interdependence. In Chap. 7, the definition of interdisciplinarity was—as one of the criteria—based on the fact that interdisciplinarity is a response to questions, problems or issues (Szostak et al., 2016). That also indicates that knowledge, procedures, methods, and collaboration will always be within a new context with a new purpose and therefore Bronstein's criteria for success are rather important.

The goal of a study based on Bronstein is to distinguish the necessary competencies from those currently in use. It identified 38 competencies or traits in interdisciplinary collaboration, ranging from content to generic competencies such as leadership and teamwork (Brandstädter & Sonntag, 2016). There are overlaps in the meaning and naming of the competencies but still we can see that the lists of competencies, including both generic competencies and traits, are increasing. What is clear is that several studies find that traits and personal qualities, such as commitment, mental readiness, helpfulness, reliability, openness, and appreciation, etc., all play an important role (Holgaard et al., 2025).

As reported in Chap. 6, the lists of employability competencies also seem to be increasing. Skills and competencies are not fixed 'products' of learning; how they unfold and are used in practice depends on one's mindset. Students can learn some skills and competencies in education; however, to avoid overloaded curricula, they should learn to adjust their knowledge and competencies to new problems and situations. Instead of integrating more competencies and traits in education, my recommendation would be to focus on some core competencies combined with the training of transfer and transformation. Students should learn generic competencies that are applicable to both disciplinary and interdisciplinary contexts, focusing on methods for approaching new situations.

8.2 Boundary Crossing as a New Core?

The PBL communities have the narrative that students learn methods and methodologies, and that deep learning in the projects would compensate for the lack of knowledge from taught courses, as the students learn to transfer knowledge, methods, and generic competencies from one project to another. Maybe that is right since PBL engineering graduates seem to work effectively at work; however, it could also be improved.

Disciplines form constructions of knowledge, learning, and culture, which saturate the discourses, the approach to understanding societal problems, and the organization of learning.

As discussed in Chap. 7, interdisciplinary competencies have better conditions for being achieved in interdisciplinary teams across the disciplines rather than in new interdisciplinary programs, which often "become new disciplines". Crossing disciplines, the students have their disciplinary knowledge in their luggage, and they have certain ways of collaborating and approaching societal problems. This is the situation in work life where graduates have to work across disciplinary boundaries and often in a transdisciplinary context across educational backgrounds. The graduated engineer will have to work and navigate through many unknown situations.

Boundary crossing might therefore be one of the core competencies that could merge many other competencies and traits. Fortuin et al. argue that boundary crossing is a generic academic competency and could be a new core competency that is feasible to integrate into engineering education (Fortuin et al., 2024). I can see the potential for this.

According to Akkerman and Bakker, boundaries are defined by differences in sociocultural practices, actions, or interactions (Akkerman & Bakker, 2011a, 2011b). In a banal way, boundaries are the difference between what is known and unknown, or unfamiliar practices. This can be knowledge, methodologies, organization, collaboration patterns, communication, values, or basic procedures. For example, the first month of a new job will be incredibly tense as procedures, culture and job function will have to be figured out. You will of course bring experiences and functional understanding of the procedures; however, the way it is carried out might be different.

Boundaries illustrate not only differences but also types of sameness; there must be a logic or a purpose to combine two fields, otherwise it would not make sense. Alternatively, interdisciplinary collaboration is prompted by new social and scientific problems which form the sensemaking.

Boundary crossing can be facilitated by boundary objects, which in turn facilitate the boundary-crossing process, and in the literature they are most often referred to as types of artifacts. Leigh Star defines boundary objects as something people act with and towards, and can be theories as well as more practical artifacts such as portfolios, etc. (Leigh Star, 2010, p. 603). In interdisciplinary collaboration, the problems can be regarded as types of boundary objects for those people working on analyzing and solving them.

Boundary crossing is not only relevant in academia. There are various types of boundaries, including cultural, institutional, and organizational ones. At least the transition from education to work will be a boundary-crossing process. Collaboration between two individuals can feel like a boundary when one is unaware of the other person's knowledge. In general, as specialization increases, so does the need for boundary crossing. Many authors also draw perspectives on intercultural learning, where there are boundaries between, e.g., Asian and Western cultures. However, learning to cross disciplinary boundaries might not automatically facilitate intercultural boundaries. Intercultural boundaries are much more complex in contrast to disciplinary boundaries which share scientific thinking.

Project work in engineering education can allow students to learn to cross transdisciplinary boundaries if the challenges stem from societal practices outside of academia. Students then can have the option to collaborate with individuals from a company, a social institution, a network, or other organizations that are also outside of academia. A study from South Africa integrates activity theory and service learning, with students conducting the boundary crossing and educators as boundary workers or brokers facilitating the students' work, giving new insights on the roles of educators as facilitators of transformation (McMillan, 2011).

8.3 Learning Mechanisms at the Boundary

In a review on boundary crossing and boundary objects, Akkerman and Bakker emphasize the learning potential of boundaries as dialogical phenomena and reveal four main clusters of learning mechanisms: identification, coordination, reflection, and transformation (Akkerman & Bakker, 2011a) (Fig. 8.1). It is these four stages in learning that Fortuin et al. find possible to integrate into engineering education to strengthen the competency development and to develop a methodology for learning boundary crossing (Fortuin et al., 2024).

Boundary crossing is typically viewed as an individual rather than an organizational activity. The identifying process encourages othering and legitimate coexistence. Othering is the process of defining one's own activities in relation to those of others. This is the key to variation: you learn about your practice and values by reflecting on what it isn't. These are also findings from research on interdisciplinary learning where students gain a deeper understanding of their discipline by comparing it to different approaches (Routhe et al., 2023). The other process discovered in the literature by Akkerman and Bakker is a legitimate coexistence, which occurs

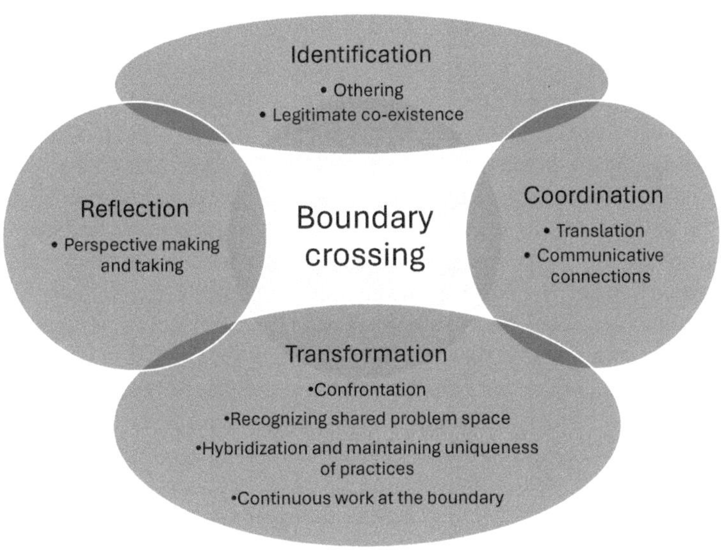

Fig. 8.1 Learning mechanism in boundary crossing identified by Akkerman & Bakker, 2011a, 2011b

when there are parallel practices that cannot combine but can operate together through acceptance and acknowledgement.

This first category of identification is vital in all sorts of interdisciplinary learning, since knowing who you are is essential and this identification process is frequently influenced by comparing one's identity to that of others. Fortuin et al. characterize the first category as assumptions, norms, and values that can pervade practice. However, the identification process may be more comprehensive because boundary crossing occurs throughout practice, typically with tacit features, as discussed in Chap. 5.

The second category, coordination, focuses on generic competencies such as communicative connections, translation, boundary permeability, and routinization. Communicative connections are intended as tools or communication structures to assist communication, although translation is far more complicated. Translation is about the meaning of actions, concepts, procedures, routines, etc., and it almost requires an insider to understand why activities are done and discussed the way they are. In Chap. 7, I suggested that interdisciplinary competencies are better learned by crossing disciplinary boundaries rather than developing new programs. However, in terms of translation, there are grounds for developing interdisciplinary programs to educate translators or boundary workers who aid in facilitating the understanding across boundaries. Routinization is the process of establishing procedures, whereas boundary permeability is about change without major obstacles and without opposition.

The third category, reflection, includes both perspective-making, perspective-taking and personal reflection. The aim is to expand sets of perspectives to look at the surrounding world. This is different from the identification processes that focus on existing practices, whereas reflection on principles should contribute towards forming future practices.

The fourth category is transformation. Transformation leads to profound changes in practices, potentially even the creation of a new, in-between practice, sometimes called a boundary practice (Akkerman & Bakker, 2011a, p. 146). This last category is the most significant. In the transformation literature, several components are mentioned: confrontation, recognizing a shared problem space, hybridization and maintaining the uniqueness of practices, and continuous work at the boundary. The first one, confrontation, sounds drastic in English, as this includes conflicts and can be interpreted as hostile. The word 'confrontation' in other languages can have a less drastic meaning, but Akkerman and Bakker describe it as a lack of realized common problems, discontinuities, disruptions, etc. The important point is that there is somehow an unsolved situation that calls for solutions and that forms a starting point for the transformation process. Maybe it is appropriate to call it confrontation so as to inform those involved that this is a complicated and even a complex or chaotic situation they will have to deal with when meeting new boundaries.

A shared problem space is more neutral and represents what Bronstein advocates for in the interdisciplinary competency framework, as well as what Carlile identifies as the first phase for transferring, translating, and transforming processes (L. R. Bronstein, 2003; Carlile, 2004). It is also what corresponds to PBL and begins with an identification and analysis of problems before the solution phases. However, understanding the problems that span across different disciplines or boundaries presents a separate challenge. In a broad interdisciplinary project team consisting of students from biology, social science, philosophy, environmental management, and electronics, it could be difficult to negotiate the problem and agree on a common definition.

Hybridization processes in collaborative work include common understandings and integrations of practices, concepts, and cultures. Like many other concepts, hybridization is a concept from chemistry, and in learning it means a process of integration and merging to create something new, which indeed is the whole idea with transformation.

Furthermore, Akkerman and Bakker find two other characteristics in the literature on transformation that maintain the uniqueness of practices and continuous work at the boundary which have to do with cultivating and keeping the boundaries.

Akkerman and Bakker regard their contribution to boundary crossings as a dialogical phenomenon. In my interpretation, this means that we can discuss the phenomenon as a process of meaning-making and identity formation that occurs through interactions with other people or various types of interactions. As learners, we are in a constant inner dialogue with our outside world, and this dialogue is diverse. I do not think we should regard these learning mechanisms as phases (Fig. 8.1). Depending on the single learner, he or she will start at different entry points and perhaps have activities that can be characterized as both identification and reflection. I see these four mechanisms as different ways to understand our practices and act at the boundaries.

In pragmatic terms, however, this categorization might be advantageous, and boundary crossing will not always involve a transformation. Perhaps the identification is sufficient, or maybe the coordination will be enough if it is related to interdisciplinary boundary work in education. These four main learning approaches also illustrate that boundary work is highly complex and can be interpreted and analyzed in many ways. Coordination is one of the most accessible mechanisms, as this will contain elements of project management and leadership, whereas identification, reflection, and transformation indeed are embracing actions, behavior, and values.

Fortuin et al. provide a more general interpretation of the four main learning processes (Fortuin et al., 2024). This is necessary if boundary crossing should become part of the engineering curriculum, as there is a need for simplifying the complexity. On the other hand, we might jeopardize the concepts, as transformation might include change in identity, values and behavior. I am not sure if this level of transformation can be achieved in one course or even three courses.

I do not see transformation as the same as transformative learning. Transformative learning, as explained in Chap. 5, concerns the development of identity, values, and attitudes that will be a significant effort and will take time. I understand transformation as a more task-oriented activity, whereas transformative learning is in inner individual learning process. Transformation is the change in practices, skills and competencies, although it is with both emotional and cognitive involvement and will have overlap with what is understood as transformative learning. Activities are done by individuals, and this is hard to separate.

The societal system determines the behavior and values among the graduates if they want to have a career and income to raise families. Somehow the system seems to swallow individuals. There are so many dependencies and constraints for both students and graduates that boundary crossing as a type of transformation is not just learned in education, but the learning can be initiated.

The awareness of practices and learning of how to develop boundary-crossing competencies can be integrated as methods in line with the understanding of generic competencies development as methods, and we can start to talk about it. There is an issue that educational research somehow always has to show its worth within engineering education, often resulting in simplification and pragmatism. With this risk of being too pragmatic, I will unfold generic competencies as methods in the next part.

8.4 Transfer and Transformation as Generic Competencies

We address transfer and transformation in the book *Reshaping Engineering Education* (Habbal et al., 2024). We distinguish between transfer and transformation of learning with transfer occurring between similar contexts and transformation between unlikely settings. This can be observed by how experience and information from one situation can be modified and applied in another. We can help our students to become better at this through education, and one suggestion for facilitators is to ask students to explore comparable situations or scenarios, or very different situations, and then train the students to think in parallel examples.

Transfer indicates that you can apply learned knowledge and competencies in similar situations, whereas transformation involves the ability to recognize and apply knowledge, skills, and competencies in new and different situations (Fig. 8.2). For example, if a person learns how to solve a particular type of problem in math class, they should be able to apply that knowledge to solve similar problems in similar contexts. This is very much the way to learn math by solving a series of smaller assignments with a change of one or two variables in the formulation of the assignment. This is an example of transfer.

Transformation involves the ability to adapt, modify and integrate existing knowledge, skills, and competencies to meet new demands. For example, if a team has developed strong problem-solving methods in sustainable planning of an energy supply system in a smaller city, they may be able to apply the design and analysis methods to designing a combined energy and transport system between two major cities. They need to modify or adapt their knowledge to be effective in the new context. In other words, transformation of competencies is about adapting and modifying existing knowledge and skills to be efficient in new and different contexts with different parameters.

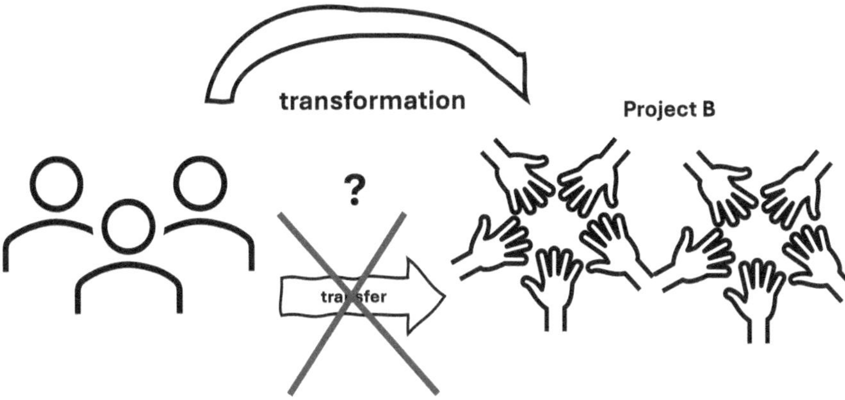

Fig. 8.2 Illustration of transfer and transformation between two contexts (Habbal et al., 2024)

Transformation in a boundary crossing context is about crossing the boundaries between two different practices (Akkerman & Bakker, 2011a). This will involve applying knowledge and competencies learned in one situation in a new situation and the individual capacity to transfer or transform one's knowledge is a must in boundary crossing. Therefore, the boundaries require transformation competencies at the individual level, see Fig. 8.2.

Both the transfer and transformation of skills and competencies are important for success in today's rapidly changing world, where individuals are often required to apply their knowledge, skills and competencies in new and different contexts. However, while the transfer of skills can be relatively straightforward, transformation can be more complex and may require a deeper understanding of how skills are applied in different contexts.

When studying interdisciplinary projects, we identified that students had significant difficulties in transferring and transforming their generic competencies learned in the disciplinary project context to a broad interdisciplinary project (Kolmos et al., 2024; Routhe et al., 2022). The lack of ability in applying simple project phases, problem analysis approaches, collaboration and communication techniques in a broader and shorter project team was surprising for us. They had no methods for transformation, and they thought they had to start all over again.

Students learn to collaborate in project teams, and when they move from one project team to the next within their disciplines, they might speak the same language and continue to exercise the same skills. They have favorable conditions for transferring their acquired competencies. However, a whole new problem arises when students are suddenly required to work in interdisciplinary teams and are unable to understand each other's information. Students will need to build on their previous experiences while also adapting to the current situation. Therefore, learning to collaborate should be combined with the methods or even methodologies of transferring and transforming.

So, what does this mean? It means that students need to learn to decode and analyze situations by identifying the problem, purpose, the involved actors, their collaborators or partners, and coordination needs. They do not just need to learn to do collaboration or do project management; they need to learn collaboration and project management in a context and what factor(s) influences the doing. To learn to analyze the relation between doing and the contexts. Some will indicate that this is reflection, others that we are dealing with meta-competencies.

Carlile (2004) defines different boundary processes as transfer, translation, and transformation in relation to innovation. In his perspective, transfer is a process applied for known knowledge areas, whereas transformation is a process with increased novelty. In between transfer and transformation, he has a level of translation that is a semantic approach of being able to understand one another (Carlile, 2004). A higher degree of novelty will influence all variables and involve setting up new organizations. Indeed, what he is indicating is that transformation also includes the ability to create a new organization.

8.5 Transformation and Meta-Competencies

Meta-competencies address the dichotomy between transfer and transformation. You can transfer the skills you have from one situation to another, but as indicated, it will be more difficult if the settings are not identical. The difficulties that students have in transferring generic competencies do not have to be in a shift from disciplinary to interdisciplinary; it could just be a new situation with a variety of components. It is analogous to Vygotsky's zone of proximal learning in that being outside the zone makes it difficult to connect with your inner cognitive framework (Vygotsky, 1978). Or we can compare to Ference Marton's concepts of sameness and difference in transfer, where you actually need a difference in order to learn — not too different; however, as there should also be a sameness for recognition (Marton, 2006).

According to Dohn, the transfer concept is part of both the learning theories of behaviorism and cognitivism. Behaviorism focuses on stimulation of recalls of knowledge and behavior through similarities, just as the example of Pavlov's dog. Cognitive theories find abstraction as a part of transfer and application of knowledge in new situations. This is based on identifying shared knowledge structures, e.g. as the logic in mathematics that can be applied in new situations (Dohn & Markauskaite, 2020; Dohn et al., 2020). Dohn et al. continue using the concept of transfer when they look at situated cognition and co-creation. I will argue that this is misleading as the learner has to be aware of their own learning processes, which are different for transfer between similar situations and transformation between different situations. In such cases, routines cannot just be repeated.

For cognitive learning theories, the abstraction process leads to the concept of meta-competencies which are the higher-order abilities that have to do with being able to learn, adapt, anticipate, and create, rather than being able to demonstrate that one has the ability to do so (Brown, 1993, p. 32). Brown argues that there are such long lists of competencies to be fulfilled that the core of the whole competency development is for students to develop the relevant knowledge and skills to carry out the work. Meta-competencies are part of the transformation and learning process along with the ability to apply and reflect on the effectiveness of the competencies in use (Brown & McCartney, 1995).

Meta-competencies are the abilities to generate competencies, and metacognition, a related concept, is the act of learning about how to learn (Brown & McCartney, 1995). A competency provides methods for putting knowledge into practice. Meta-competency includes strategies for dealing with competencies in action, as competency within a specific field transforms through interaction with other competencies. Competency and meta-competency have a dialectical relationship, and their borders can be blurred. Meta-competencies are defined as the development, adjustment, and use of competencies at a higher level of reflection (Brown, 1993). As meta-competencies are about developing competencies, I would recommend that the development of competencies should be integrated with the learning of competencies.

Studies indicate that project-based learning gives students the possibility to learn transformation (Safitri, 2024; Von Kotze & Cooper, 2000). Von Kotze and Cooper (2000) focus on transformation for adult students and conclude that the challenge is how we apply PBL to build up the learners' confidence, which is an essential part of learning.

In chaotic situations, people will have to manage transformation fast, which involves generic competencies, experiences, knowledge, and very different collaboration partners. Chaos indicates that there is no time to analyze or reflect but rather the need to act and react. It also means that students need to have experienced the activities of acting and reacting in unfamiliar situations because experiences form a first confidence of acting in chaos.

8.6 Transformation Competency to Handle Chaos

As transformation is a process in learning where you activate past experiences, knowledge, skills and attitudes, it should also be possible to learn as a method or even as a methodology more than an exact skill or competency. This actually counts for all generic competencies since graduates will not usually stay in one knowledge domain for the rest of their lives. The individual must be able to develop his or her own professional and organizational competence, i.e. both to assimilate in relation to known frameworks but also to accumulate and transform learning from one context to the next and from one conceptual understanding to a new one. Accumulating and transforming knowledge and competencies is also about being able to choose strategies, methods and techniques for specific situations including how a personal learning strategy is built and developed. Individually as well as collaboratively, it is important that the student is able to continuously upgrade and develop his or her knowledge.

The competency to be learned, therefore, is how to master one's own learning processes and be able to participate in diverse organizational settings. In Chap. 2, I referred Snowden and Boone's understanding of chaos and what it takes to be able to handle it (Kurtz & Snowden, 2003). For the simple, complicated, and complex domains, the first phase of the learning process is to sense in various formats. For the simple domain, you have known categories by which you can analyze a situation. For the complicated domain, sensing is followed by analyzing, often across disciplinary boundaries in a system perspective, which is manageable. For the complex domain, the problems and interactions among knowledge domains are not obvious, and therefore it is a process of probing before sensing and acting.

For chaos, we look into disaster, mess, turbulence, and most often insecurities. Chaos can occur in periods across all the other domains, e.g. in an organization or in innovation where some of the fundamental elements do not work out as expected. It can be caused by more serious issues such as climate, war, or terror, all of which create a situation of urgency. Therefore, chaos is not necessarily an isolated situation waiting for us, it already exists in current organizations and in personal lives, with conflicts, death, divorce, losing jobs, economic loss, etc. Chaos is a part of personal life and societal relationships. In boundary work across several knowledge domains, there will be periods or situations of chaos in the collaborative progress. Conflicts, misunderstandings, confrontations, lack of problem solving, threat, anxiety, hopelessness, and many more situations will occur in an effort to create common understanding.

In contrast to other fields that are dominated by simple, complicated, and complex problems, chaos inverts the order of action. In fields with simple and well-known knowledge, one will begin by identifying and reflecting on issues to gain new insights before taking action. However, in the chaos domain, one will start by acting before sensing and thinking. The initial action will often involve creating constraints to limit damage, followed by a period of reflection to assess whether these solutions are feasible or if adjustments are necessary. Reactions and new learning are then possible.

We do not teach engineering students to act in chaos. They might acquire this competency as theoretical knowledge, but they do not get to experience how they would react in this situation. We also don't teach engineering students how to manage transformation, and only a few places address boundary crossing as a competency.

In Chap. 6 when discussing generic competencies, I emphasized the learning of generic competencies by using experimentation instead of reflection (Kolb, 1984). If we start with experimentation instead of reflection, actions can lead to reflection. In addition, if experimentation is not combined with reflection, it will lead to trial and error and become tacit instead of leading to more conceptual learning.

In chaotic situations, conceptual learning might be of less importance in the first place, and it could be more meaningful to let the student experience the trial-and-error processes. This is assuming the conditions that students also learn the frameworks of chaos compared to the other problem domains, different learning circles, and learning processes to contextualize their learning in a broader perspective. However, students need to learn to react on the basis of their conceptual and scientific learning.

Learning is hard to quantify because we are all different. We are motivated by logic, emotions, and personal experiences, and students therefore should always be presented with various models. Boundary crossing and transformation skills differ from learning to act and reacting to chaos, and both can be useful. Crossing boundaries to new practices and applying competencies within new contexts involve meeting with something new and unknown.

Fortuin et al. (2024) have another example, and even though I have already mentioned that I find transformation might be a difficult level to reach, their model inspired by Akkerman and Bakker should not be ignored. This methodology can be applied more broadly in education to help students learn transformation as a generic competency. Fortuin et al. define identification as both one's own and others' assumptions and values in order to understand perspectives.

This is a really difficult process. How do you identify and understand others' expertise if the language is different? The coordination concerns the effective communication and identification of boundary objects to facilitate this. Reflection concerns how to learn from others and shift perspectives. Transformation affects how values and personal behavior change in a new practice. The authors propose a boundary-crossing learning trajectory in the curriculum that will involve several courses and projects throughout the curriculum (Fortuin et al., 2024, p. 219).

Activity theories also need to be mentioned when writing about boundaries. An activity theory is a framework to understand and analyze the learning practices with interactions between individuals and their surroundings. Tools, rules, and social contexts facilitate this interaction, enabling us to understand human interactions. Each activity system establishes its own boundaries, necessitating the crossing of these practice boundaries (Engeström, 2001). Each element in the system is related to all the other elements in the system. For the subject, the learner uses tools or instruments within a community with rules and division of labor, which all leads to an object—the outcome.

The concept of expansive learning involves the transformation of an activity system and is understood as a form of collective learning (Engeström, 2001). Expansive learning is a process starting from questioning and experiencing contradictions to modeling, implementing them, and consolidating new practices. This is a kind of organizational learning process. The single learner moves from being an isolated individual participating in the collective activities to someone transforming the work. There is an embedded agency for the individual to interact with the system and combine the individual's inner motivations with actions and collective actions.

Activity theory has undergone at least four generations (Engeström & Sannino, 2021). In the third generation of activity theory, knowledge knots are identified. The title of Engeström's book *From Teams to Knots* describes a dynamic system that is constantly developing, where stable teams do not exist and individuals work within organizations characterized by temporary relationships. This is the organization pattern that the engineers describe in the InterPBL studies on engineering practice, where they work in parallel projects and contribute their expertise (Friedrichsen et al., 2024; Holgaard et al., 2025). The fourth generation of activity theories embraces uncertainties and fluctuating organizations where it might be difficult just to identify the involved actors and not least the objects or purposes of actions with shifting actors. The reflections on the development of activity theory does not include the factor of chaos; however, the fluctuating processes might create moments of chaos.

Table 8.2 gives an overview of the various theories and the last method mentioned in this chapter is the one of project phases. The project's first phase is to identify and analyze the problem, which will be the project's core and the focus of the students' reflections throughout the process. The project phases in PBL are related to the scientific content; however, the students are learning to identify problems, situations, involved actors, and contradicting interests, and based on the initiating analyses they learn to formulate problems and hypotheses.

This can be a tremendous advantage in relation to decoding and understanding chaotic situations. In a chaotic situation, the initial reaction should be to try to decode what it is about. A decoding process might be facilitated by the learning of problem analyses both from a cognitive and content perspective and by the generic competencies, including transfer and transformation.

8.7 Emotions as Part of Transformation and Chaos

The emotional aspect has never been a component of PBL competencies, nor is it included in the core principles of boundary work and transformation. In collaborative learning, conflicts are unavoidable and conflict management is indeed emotional. Perhaps this is also where the existing student-centered learning methodologies have potential for development. Why did we not include the emotional part more explicitly?

To be honest, I have never doubted the role of emotions in the learning of competencies, which was part of my first activities as a PhD student (Kolmos, 1989, 1990). Asking for money to improve students' emotional collaboration was not valid throughout my lifetime and we would not be able to do so anyway. Around 2000, I recall one of my first PhD students requesting that a meditation course should be recognized as a PhD course because it had inspired him to learn so much, to which I promptly declined. We were establishing the recognition and prestige of something new—engineering education research—with logic and rationalism at the forefront of the discourse. This was the only way to communicate with the deans and managers at the time. Today, it is different and the #MeToo movement has given people permission to communicate and talk about emotions.

The learning of transformation and generic competencies relies on emotional experiences as well as cognitive and emotional reflection. Within engineering education research, there is a movement on emotions that witnesses the shift in time and

Table 8.2 Overview of key learning elements in boundary crossing, reaction to chaos learning and problem analyses

(Akkerman & Bakker, 2011a, 2011b)	(Snowden & Boone, 2007) chaos	(Kolb, 1984)	(Fortuin et al., 2024)	(Engeström, 2001)	(Algreen-Ussing & Fruensgård, 2006) Project phases
Identification: • Othering • Legitimate co-existence Coordination: • Translation • Communicative connections Reflection: • Perspective making and taking Transformation: • Confrontation • Recognizing shared problem space • Hybridization and maintaining uniqueness of practices • Continuous work at the boundary	Act Sense Respond	Experimentation Experience Reflection Conceptualization	Identification Coordination Reflection Transformation	Primary contradiction and questioning Secondary contradictions Modelling new solutions New models Implementation of the new model Reflection Consolidating new practices	Initiating problem Problem analyses Definition of the problem Problem-solving methods Demarcation Specification of requirements Solving the problem Implementation (prototype) Evaluation and reflection

acceptance. In 2025, the European Journal of Engineering Education published the first special issue on emotions in engineering education. However, research on emotions is much broader and has become a trend in educational research generally.

The relationship between emotions, ethics, and sustainability is addressed in the engineering education research field. The mainstream approach in ethics research is based on cognitive and rational dimensions. In a recent study, the findings revealed that emotions like embarrassment and anger were focal points in students' reactions to ethical cases together with feelings of powerlessness (Kotluk & Tormey, 2025).

Sustainability also comprises emotional challenges for students as this field is characterized by open-ended problems, with Holmén and Lönngren identifying different types of emotional reactions to these types of situations (Holmén & Lönngren, 2025). Holmén and Lönngren call for emotional scaffolding as a system to be used for educational design and evaluation and point at two dimensions that can form four fields: high versus low activation of emotions and negative versus positive emotions. Different situations such as success and failures can create very different feelings in that model.

As such, these can be important models for students to start reflecting their emotional reactions in the project process. More open-ended problems and challenges create frustration among students, something that I think most facilitators will recognize. However, it is important to emphasize that frustration can be an important emotional stage in any transformation process and can be used positively in creating new solutions. The feeling of frustration will occur many times in life, and a chaotic situation will create frustration.

Ippolito has studied emotional processes in team projects and based on the data, she gives four recommendations for project design: to be clear on goals, increase student control in assessment, communicate on the subject of about emotions, and train teachers in talking about emotions (Ippolito, 2025). These are all very good proposals if the emotional argument is not used to give more narrowly scoped problems and projects to students.

Studies in programming also identify emotional reactions both in successful learning situations and in cases where it is difficult to have more negative feelings (Atiq & Batra, 2025). Indeed, emotions are part of learning and there will always be ups and downs in all types of learning environments. Teachers can influence students' experiences of the learning environment and factors as enthusiasm and encouragement in various learning methods are important (Hartikainen et al., 2022).

In the transfer and transformation literature there is an increasing focus on the emotional part. Emotions are contagious among both individuals and teams (Goldenberg et al., 2014). There are some underlying mechanisms, which can be hard to identify and can suddenly dominate a culture. Who has not experienced a time when one class was very positive and another negative in their evaluation despite the fact the content was the same and presented in the same way. Emotions are contagious and meeting other people's feelings and reactions might create similar feelings, and when something might be experienced as hard or problematic, it might taint one's own experiences and approaches (Parkinson, 2011). Emotions are also periodic; they occur and disappear again in the transfer and transformation processes (Nemirovsky, 2011).

Regardless of negative or positive feelings, emotions do exist for shorter or longer periods and they also play a role in activating the previous experiences and knowledge being transferred or transformed. In a review by McConnell and Eva (2012), the findings support the fact that positive feelings might increase the transfer of knowledge and skills, whereas negative feelings might hinder it (McConnell & Eva, 2012). One of the conclusions is:

> The literature reviewed in this article suggests that when students perceive a learning situation as threatening or frightening, they may have better memory of the emotional event because of cognitive rehearsal and elaboration, but they may be less likely to make broader connections, thereby precluding transfer of the learned information to other clinical settings. Therefore, although pimping may enhance memory for certain situations, creating the impression of influential teaching/learning, it may actually impede the application of knowledge gained to new contexts. (McConnell & Eva, 2012, p. 1321)

There is no formula for how and if emotions would enhance transfer and transformation. I dare to conclude the same in relation to chaos, where the reactivation of knowledge, skills, and competencies is urgent in the immediate reactions to the situation. One emotional reaction could be to entirely freeze, for example with test anxiety; however, the opposite can also happen.

8.8 Agency and Resilience

Emotional reflection should be among the strategies for increasing students' resilience. Emotions can have negative and positive effects on learning, and the most important part is to have a dialogue about one's own emotions and talking about them in education. When asking students to reflect, they should consider their own feelings and experiences not just their activities and knowledge. Emotions need to become more present in the daily discourses in education.

In Chap. 5 I touched upon a growth mindset and mastery that relates to resilience. If you have a high level of resilience, it can overlap with having a growth mindset and the belief that you can develop your talent, knowledge, skills, and competences, and that you can master difficult learning situations and take risks.

Several engineering education researchers have applied the framework of fixed and growth mindsets for understanding the behavior, learning, and curriculum interventions of engineering students. In a US study, it is proven that students can shift their mindset from fixed to growth, and without this, the educational system would have problems since a growth mindset indicates learning (Frary, 2018).

Another study indicates that the retention of engineering students involves a type of growth mindset, and that curriculum and pedagogical approaches are critical in developing the mindset and learning growth of students (Campbell et al., 2021). Having a positive and growth mindset will impact learning outcomes and competency development (Lottero-Perdue & Lachapelle, 2020; Richardson et al., 2021).

Studies on interdisciplinary competencies and mindsets relate to critical thinking for students and they see the interdisciplinary conversation as a means of achieving critical thinking (Ho et al., 2023). Learning various viewpoints and perspectives fosters the development of a broader and more critical mindset. Another study shows similar findings in relating mindset to intelligence or smartness and beliefs in engineering education in an interdisciplinary context. The interdisciplinary context opens up students' beliefs (Dringenberg et al., 2019).

Having a growth mindset is also aligned with the understanding of agency, which is another concept that unfolds diverse meanings in different contexts (Emirbayer & Mische, 1998; Hitlin & Elder Jr, 2007). In educational research, the concept is related to self-efficacy and active learning as the learners show agency in constructing knowledge (Eteläpelto et al., 2013). This involves having the capacity to act and the ability to direct one's own learning, which is a fundamental criterion for PBL and other active learning theories in general.

Research also indicates that PBL can have a positive impact on the development of agency (Wong, 2023), and this is a condition for learning practices within a PBL learning environment. The research literature on PBL and agency is not overwhelming, although studies on agency in relation to sustainability in a PBL learning environment indicate a gap between knowledge and actions (Guerra et al., 2022).

It is obvious to draw parallels to the relationship with chaos. Agency in this context will involve being able to respond to chaotic situations no matter if they are periodic, dramatic, or a nature of general collapse. This also raises the question of the emotional impact on agency. This will lead to a deeper psychological study that is beyond my scope in this book, and it is not clear if such studies would even lead to any clear conclusions. I am convinced that we, as humans and learners, are influenced by so many factors that even negative emotional self-perceptions can be advantageous in collaborative actions that minimize chaotic situations.

Resilience is also a concept that has been discussed to be more actively integrated into engineering education (Davidsen et al., 2019). It is another concept that is difficult to define and is related to identity, persistence, retention, self-efficacy, student engagement, and student success (Hunsu et al., 2021). It could be an overlapping concept. A recent study focuses on engineering students' understanding of resilience (Wint & Direito, 2024). The findings indicate that students found resilience important; however, they did experience positive support from the educational system.

A review study on resilience in engineering education found that the concept was linked to both engineering students and the description of the content in teaching (Winkens & Leicht-Scholten, 2023). Other research studies have reached similar findings, indicating that while resilience is regarded as a desirable trait for engineering graduates, academic staff do not take it into serious consideration in their teaching or course design (Winkens & Leicht-Scholten, 2023; Winkens & Leicht-Scholten, 2022; Wint & Direito, 2024). A study among nursing students suggests more emphasis is needed on engaging students in reflection, and that resilience can be encouraged by greater self-assessment and reflections on problems in collaboration (Chen, 2011).

I would strongly recommend that psychological elements become part of a curriculum so that students have frameworks and concepts to understand their own emotional reactions and ways to move from knowledge, experiences, and emotional stages to actions and agency. This is not seen as an easy fix. Working on one's own identity, and not at least on reading and understanding others' responses and perspectives, could be painful and experienced as damaging; on the other hand, life is sometimes painful, and there is a need for deconstruction before reconstruction and new hopes.

8.9 A Note on Well-being

I will briefly address 'well-being' as a field that has become popular. The understanding of well-being differs as every other concept does. The book *Wellbeing: The Five Essential Elements* defines wellbeing with five characteristics for citizens in general: career, social, financial, physical, and community wellbeing (Rath & Harter, 2010). These variables are very close to the happiness variable, but the book clearly states that this is not about being happy or rich, but about how you live your life. It is about commitment, control of one's own life, meeting challenges, being connected to the community, and having confidence (Geldhof et al., 2015).

In a US study on PBL and well-being, academic staff find that the collaborative and social environment has a positive impact on students' well-being (Rubino, 2024). A study on the same subject found that factors of personal values, agentic action, interactions within the learning environment, and external support have had a positive impact on the students' academic well-being (Chen et al., 2023). In another study, it was found that running PBL for a shorter period still gave students the experiences of taking responsibility, coping with challenges, and building relationships that last after the project period (Allison et al., 2015). In many ways, this is what we expect from the PBL system—that it gives students a better sense of not only a safer environment in the project teams but the ability to develop their own agency.

There is, however, the other side as shown in an article documenting social environments as being a constraint on students. In their study, the authors interviewed four men returning to the educational system and their struggle in both the social environment and the professional life (Servant-Miklos et al., 2021). Their hopes of becoming engineers with a status in society and resilience from their past life were the drivers for remaining in the study.

PBL can be great, but we must not forget that the social environment can also create issues for some students. For example, for introverted students who have to manage the hurdle of becoming social and learning to interact with other students, and for female students in engineering who have to adjust in teams who spend a lot of time together with men's culture and jokes dominating their group rooms. It can also be returners to the educational system who have families and suddenly spend a lot of time together with primarily younger male students. There are conflicts and different interests that can be hard to deal with, and it is not automatically a safe environment. But for a time it might be the place where they have to learn to cope, manage conflicts, and develop new strategies for collaboration.

It is important to help students develop their own personal and collaborative strategies and acknowledge that the social environment might not always be a safe one. We have emotional imbalances; at times everything seems within reach, and a moment later obstacles appear to pile up, making any progress difficult. As with modelling for resilience, there are risks and potentials in a constant dialectic process with life and learning progressing in ups and downs, which can be difficult to accept and recognize. We are pushed around in an arena, from being content to being concerned, almost depressed, and demotivated.

When you experience emotional instability, you need to temporarily re-establish your grounding and balance. It is a balancing act, comparable to a pendulum that swings back and forth until it finds its equilibrium point. To frame emotional imbalances, experiences must be paired with various forms of reflection and understanding. The reflection and development of strategies help to further develop resilience and give an overview of what can be done in different situations. This ought to be part of the reflection and development of generic competencies as it develops students' resilience. Learning from failures, conflicts, misconceptions, etc. can be some of the most important learning situations.

There is a risk that if well-being is combined with the values of the #MeToo movement, it might result in values that contradict PBL. The #MeToo flavor has been integrated into human resource systems and concerns broader aspects such as communication and the risk that we lose the learning potential in e.g. othering and confrontation. In a PBL culture, students are required to learn how to be responsible for their actions and to communicate socially in project teams. This includes the ability to resolve and overcome issues and to engage in uncomfortable conversations. There will be plenty of conflicts, and students need to learn to solve these. They need to become resilient human engineers, capable of controlling and resolving technical and societal emergencies before they become chaotic and potentially collapse.

For this, we need an expanded curriculum with a core concentration on technical and contextual knowledge, as well as an integrated focus on learning generic competencies as methods with the inclusion of emotional factors. The methods for acquiring generic competencies must focus on the integration of reflection and action to enable the learning of agency and collaborative agency, as well as the development of self-esteem and confidence as part of resilience.

8.10 Reflection 8: Living with Chaos

This chapter includes some of my imaginations for the future curriculum, as well as reasons for how PBL might help integrate chaos into the curriculum. I also made an argument for the chaos agenda, and potential of PBL. Many institutions that want to apply PBL today should accomplish this in a contemporary way, including curriculum flexibility and reconsideration of the outcomes and learning of generic competencies.

Students should learn to decode chaotic situations. Chaos is unexpected, and it can be so for periods of time, or it can be a real disaster like climate change. If chaos is periodic, it will normally be a part of complex systems that have become unmanageable. No matter the kind of unexpected situation, students should learn to decode and understand what constraints can limit the damage and what kind of knowledge should be necessary in this process. Knowledge can be either human or AI-generated responses, and even if the latter is the case, it must be validated in relation to the specific circumstance.

The difficulty here is speed as there will be no time to probe, sense, or analyze as in the other problem domains proposed by Snowden and Boone (2007), and everything must be done quickly. This also implies that it is beneficial to gain some experience by trying out the processes, both for the knowledge component and for collaborative and coordinated actions in education.

PBL systems will benefit students' understanding by learning how to identify, analyze, and characterize problems. This should in theory provide competencies to quickly create constraints in the early stages of potentially chaotic circumstances. Risk management might be another element which can prepare students through the focus on possible risks in technological development, and it might help in providing an analytical understanding. Risk management should be seen both in their discipline and in relation to interdisciplinary initiatives. Fast problem identification can be challenging; nevertheless, if students have encountered a variety of problem scenarios, both wide and restricted in scope, it might provide a foundation for how to tackle varied chaotic situations.

It should also be simple to integrate risk management with students' present projects. Risk management is a sub-discipline that they frequently study when working on projects by evaluating what could go wrong. In every EU or national research council application, it is required to write about risk management as part of the project process. So, even if the knowledge is generally available, there is a limited tradition of integrating risk management and evaluation in both research and education. To integrate risk management and evaluation into the curriculum and academic culture, it must first be articulated and made an obvious part of the criteria for academic applications.

In addition to the knowledge dimension there are generic competencies. Boundary crossing and transformation in particular can both be viewed as new core skills capable of connecting, relating, and developing new competency profiles across existing disciplines, organizational units, industries, and cultures. We can, however, teach generic competencies the wrong way. Collaboration should be learnt as a process and a method rather than a fixed ability. As facilitators, we should be much better at asking the question: What if we change some of the variables and what would you do then?

Besides a reform of the existing teaching and potential curricular development, these types of generic competencies are not easy to learn. Understanding the perspectives and language of one's counterpart requires not just interdisciplinary competencies but also some level of emotional growth. Emotions cannot be reduced to a formula. We do not always understand why we react the way we do, but we can start to talk about it. We can ask the students to reflect on their feelings: Did ill-organized activities and problems generate any emotional responses? Was I content with my own efforts and pleased with the reactions of my peers? How do I learn to become more active and faster in my analyses and decisions? Do I trust my peers and their knowledge? And if I don't, what are my barriers with trusting other people? How do I become a better version of myself in boundary work?

References

Akkerman, S. F., & Bakker, A. (2011a). Boundary crossing and boundary objects. *Review of Educational Research, 81*, 132–169. https://doi.org/10.3102/0034654311404435

Akkerman, S. F., & Bakker, A. (2011b). Crossing boundaries between school and work during apprenticeships. *Vocations and Learning, 5*, 153–173. https://doi.org/10.1007/s12186-011-9073-6

Algreen-Ussing, H., & Fruensgård, N. (2006). *Metode i projektarbejdet: problemorientering og gruppearbejde*. (1. udgave, 11. oplag. ed.). @Ålborg Universitetsforlag.

Allison, P., Gray, S., Sproule, J., Nash, C., Martindale, R., & Wang, J. (2015). Exploring contributions of project-based learning to health and wellbeing in secondary education. *Improving Schools, 18*(3), 207–220.

Atiq, Z., & Batra, R. (2025). A multi-modal investigation of retrospective emotions during a programming task. *European Journal of Engineering Education*, 1–33.

Beddoes, K. (2020). Interdisciplinary teamwork artefacts and practices: A typology for promoting successful teamwork in engineering education. *Australasian Journal of Engineering Education, 25*(2), 133–141.

Borrego, M., & Cutler, S. (2010). Constructive alignment of interdisciplinary graduate curriculum in engineering and science: An analysis of successful IGERT proposals. *Journal of Engineering Education, 99*, 355–369. https://doi.org/10.1002/j.2168-9830.2010.tb01068.x

Brandstädter, S., & Sonntag, K. (2016). *Interdisciplinary collaboration: How to foster the dialogue across disciplinary borders?* Advances in ergonomic design of systems, Products and Processes: Proceedings of the Annual Meeting of GfA 2015.

Bronstein, L. R. (2003). A model for interdisciplinary collaboration. *Social Work, 48*(3), 297–306.

Brown, R. B. (1993). Meta-competence: A recipe for reframing the competence debate. *Personnel Review., 22*, 25–36.

Brown, R. B., & McCartney, S. (1995). Competence is not enough: Meta-competence and accounting education. *Accounting Education, 4*(1), 43–53.

Campbell, A. L., Direito, I., & Mokhithi, M. (2021). Developing growth mindsets in engineering students: A systematic literature review of interventions. *European Journal of Engineering Education, 46*(4), 503–527.

Carlile, P. R. (2004). Transferring, translating, and transforming: An integrative framework for managing knowledge across boundaries. *Organization Science, 15*(5), 555–568.

Chen, J., Du, X., Chaaban, Y., Velmurugan, G., Lyngdorf, N. E. R., Nørgaard, B., Routhe, H. W., Hansen, S., Guerra, A., & Bertel, L. B. (2023). An exploration of sources fostering first-year engineering students' academic Well-being in a PBL environment. *IEEE Transactions on Education, 66*(5), 421–430.

Chen, J.-Y. (2011). Problem-based learning: Developing resilience in nursing students. *The Kaohsiung Journal of Medical Sciences, 27*(6), 230–233.

Claus, A. M., & Wiese, B. S. (2019). Development and test of a model of interdisciplinary competencies. *European Journal of Work and Organizational Psychology, 28*(2), 191–205.

Davidsen, J., Emtkjær, P. B., Holgaard, J. E., Therkildsen, H. F., & Telléus, P. K. K. (2019). *Study competencies: Communication, motivation, resilience, study technique and learning*.

Dohn, N. B., & Markauskaite, L. (2020). Conceptions of transfer, transformation and resituation. In *Designing for situated knowledge transformation* (pp. 39–58). Routledge.

Dohn, N. B., Markauskaite, L., & Hachmann, R. (2020). Enhancing knowledge transfer. In *Handbook of research in educational communications and technology* (pp. 73–96). Springer.

Dringenberg, E., Secules, S., & Kramer, A. (2019). *Smartness in engineering culture: An interdisciplinary dialogue*. 2019 ASEE Annual Conference & Exposition,

Emirbayer, M., & Mische, A. (1998). What is agency? *American Journal of Sociology, 103*(4), 962–1023.

Engeström, Y. (2001). Expansive learning at work: Toward an activity theoretical reconceptualization. *Journal of Education and Work, 14*(1), 133–156.

Engeström, Y., & Sannino, A. (2021). From mediated actions to heterogenous coalitions: Four generations of activity-theoretical studies of work and learning. *Mind, Culture, and Activity, 28*(1), 4–23.

Eteläpelto, A., Vähäsantanen, K., Hökkä, P., & Paloniemi, S. (2013). What is agency? Conceptualizing professional agency at work. *Educational Research Review, 10*, 45–65.

Fortuin, K., Gulikers, J. T., Post Uiterweer, N. C., Oonk, C., & Tho, C. W. (2024). Developing a boundary crossing learning trajectory: Supporting engineering students to collaborate and co-create across disciplinary, cultural and professional practices. *European Journal of Engineering Education, 49*(2), 212–235.

Frary, M. (2018). *Encouraging a growth mindset in engineering students*.

Friedrichsen, D., Holgaard, J. E., Kolmos, A., & Routhe, H. W. (2024). Surprises at work. A study of early career engineers. In *SEFI 2024 proceedings*. European Association for Engineering Education.

Geldhof, G. J., Bowers, E. P., Mueller, M. K., Napolitano, C. M., Callina, K. S., Walsh, K. J., Lerner, J. V., & Lerner, R. M. (2015). The five Cs model of positive youth development. In *Promoting positive youth development: Lessons from the 4-H study* (pp. 161–186). Springer.

Goldenberg, A., Saguy, T., & Halperin, E. (2014). How group-based emotions are shaped by collective emotions: Evidence for emotional transfer and emotional burden. *Journal of Personality and Social Psychology, 107*(4), 581–596.

Guerra, A., Jiang, D., & Du, X. (2022). Student agency for sustainability in a systemic PBL environment. *Sustainability, 14*(21), 13728.

Habbal, F., Kolmos, A., Hadgraft, R. G., Holgaard, J. E., & Reda, K. (2024). *Reshaping engineering education: Addressing complex human challenges*. Springer Nature.

Hartikainen, S., Pylväs, L., & Nokelainen, P. (2022). Engineering students' perceptions of teaching: Teacher-created atmosphere and teaching procedures as triggers of student emotions. *European Journal of Engineering Education, 47*(5), 814–832.

Hitlin, S., & Elder, G. H., Jr. (2007). Time, self, and the curiously abstract concept of agency. *Sociological Theory, 25*(2), 170–191.

Ho, Y.-R., Chen, B.-Y., Li, C.-M., & Chai, E. G.-Y. (2023). The distance between the humanities and medicine: Building a critical thinking mindset by interdisciplinary dialogue through mind mapping. *Thinking Skills and Creativity, 50*, 101420.

Holgaard, J. E., Kolmos, A., Friedrichsen, D., & Routhe, H. W. (2025). Interdisciplinarity, generic competencies, and traits in engineering practice from the perspective of early-career engineers. *European Journal of Engineering Education*, 1–17.

Holmén, J., & Lönngren, J. (2025). Expanding sustainability learning in engineering education through emotional scaffolding. *European Journal of Engineering Education*, 1–27.

Hunsu, N. J., Carnell, P. H., & Sochacka, N. W. (2021). Resilience theory and research in engineering education: What good can it do? *European Journal of Engineering Education, 46*(6), 1026–1042.

Ippolito, K. (2025). 'It's all about balance': Exploring how engineering undergraduates and teachers regulate emotions in pursuit of group-based project goals. *European Journal of Engineering Education*, 1–23.

Kolb, D. (1984). *Experiential learning: Experience as the source of learning and development* (1st ed.). Prentice Hall.

Kolmos, A. (1989). *Kvindesynsvinkler i de teknisk-naturvidenskabelige uddannelser: sammenfatning*. Aalborg Universitetscenter.

Kolmos, A. (1990). Women and men in engineering education–different approaches. In *Contributions to GASAT* (p. 298).

Kolmos, A., Holgaard, J. E., Routhe, H. W., Winther, M., & Bertel, L. (2024). Interdisciplinary project types in engineering education. *European Journal of Engineering Education*, 1–26.

Kotluk, N., & Tormey, R. (2025). Exploring the emotional dimension in engineering ethics case studies: The latent emotions and associated features. *European Journal of Engineering Education*, 1–24.

Kurtz, C. F., & Snowden, D. J. (2003). The new dynamics of strategy: Sense-making in a complex and complicated world. *IBM Systems Journal, 42*(3), 462–483.

Lattuca, L. R., Knight, D., & Bergom, I. (2013). Developing a measure of interdisciplinary competence. *International Journal of Engineering Education, 29*(3), 726–739.

Leigh Star, S. (2010). This is not a boundary object: Reflections on the origin of a concept. *Science, Technology, & Human Values, 35*(5), 601–617.

Lottero-Perdue, P. S., & Lachapelle, C. P. (2020). Engineering mindsets and learning outcomes in elementary school. *Journal of Engineering Education, 109*(4), 640–664.

Marton, F. (2006). Sameness and difference in transfer. *The Journal of the Learning Sciences, 15*(4), 499–535.

McConnell, M. M., & Eva, K. W. (2012). The role of emotion in the learning and transfer of clinical skills and knowledge. *Academic Medicine, 87*(10), 1316–1322.

McMillan, J. (2011). What happens when the university meets the community? Service learning, boundary work and boundary workers. *Teaching in Higher Education, 16*(5), 553–564.

Nemirovsky, R. (2011). Episodic feelings and transfer of learning. *The Journal of the Learning Sciences, 20*(2), 308–337.

Parkinson, B. (2011). Interpersonal emotion transfer: Contagion and social appraisal. *Social and Personality Psychology Compass, 5*(7), 428–439.

Rath, T., & Harter, J. K. (2010). *Wellbeing: The five essential elements*. Simon and Schuster.

Repko, A. F., Szostak, R., & Buchberger, M. P. (2019). *Introduction to interdisciplinary studies*. Sage Publications.

Richardson, D., Kinnear, B., Hauer, K. E., Turner, T. L., Warm, E. J., Hall, A. K., Ross, S., Thoma, B., Van Melle, E., & Collaborators, I. (2021). Growth mindset in competency-based medical education. Medical Teacher, 43(7), 751–757.

Routhe, H. W., Holgaard, J. E., & Kolmos, A. (2023). Experienced learning outcomes for interdisciplinary projects in engineering education. *IEEE Transactions on Education*, 1–13. https://doi.org/10.1109/TE.2023.3284835

Routhe, H. W., Winther, M., Nguyen, N. T., Holgaard, J. E., & Kolmos, A. (2022). Challenges for engineering students working with authentic complex problems. In *SEFI 2022: Annual conference 19–22 September 2022*.

Rubino, S. (2024). *Project-based learning and its impact on student engagement, Well-being, and learning outcomes: A study of teachers' perspectives*. Arkansas State University.

Safitri, I. (2024). Transformation of student science achievement through project-based learning (PjBL): A meta analysis study. *Scientiae Educatia, 13*(2), 146.

Servant-Miklos, V. F., Dewar, E. F., & Bøgelund, P. (2021). I started this, and I will end this': A phenomenological investigation of blue collar men undertaking engineering education as mature students. *European Journal of Engineering Education, 46*(2), 287–301.

Snowden, D. J., & Boone, M. E. (2007). A leader's framework for decision making. *Harvard Business Review, 85*(11), 68.

Szostak, R., Gnoli, C., & López-Huertas, M. (2016). *Interdisciplinary knowledge organization*. Springer.

Tripp, B., & Shortlidge, E. E. (2019). A framework to guide undergraduate education in interdisciplinary science. *CBE Life Sciences Education, 18*(2), es3. https://doi.org/10.1187/cbe.18-11-0226

Von Kotze, A., & Cooper, L. (2000). Exploring the transformative potential of project-based learning in university adult education. *Studies in the Education of Adults, 32*(2), 212–228.

Vygotsky, L. (1978). Interaction between learning and development (M. Cole, Trans.). In *Mind in society* (pp. 79–91). Harvard University Press.

Winkens, A.-K., & Leicht-Scholten, C. (2023). Does engineering education research address resilience and if so, how?—A systematic literature review. *European Journal of Engineering Education, 48*(2), 221–239. https://doi.org/10.1080/03043797.2023.2171852

Winkens, A.-K., & Leicht-Scholten, C. (2022). *Local resilience strategies for COVID19–a PBL engineering case study*. Proceedings of the 18th international CDIO conference,

Wint, N., & Direito, I. (2024). (re) defining resilience: How engineering educators understand and teach resilience. *Studies in Engineering Education, 5*(2), 175–195.

Wong, D. S.-S. (2023). *What is student agency and how might it be influenced by project-based learning?* Macquarie University.

Chapter 9
Educational Change and Transformation

9.1 Introduction

> *While all transition involves change, not all change results in transition. Changes can occur within a single historical epoch that do not profoundly affect it in any way* (Freire, 1973, p. 6).

Change is not equal to transformation. In a publication with Roger Hadgraft and Jette Egelund Holgaard, we wrote about different strategies for change: add-on, integration, and rebuilding (Kolmos et al., 2016). This publication on strategies is about moving to a more sustainable curriculum, and the framework we presented is based on Sterling's tremendous impact on the conceptualization of sustainable education. Sterling and Orr emphasize in their book:

> By contrast, second order change and learning involves critically reflective learning, when we examine the assumptions that influence first-order learning. This is sometimes called 'learning about learning' or 'thinking about our thinking'. At a deeper level still, when third-order learning happens we are able to see things differently. It is creative, and involves a deep awareness of alternative worldviews and ways of doing things. It is, as Einstein suggests, a shift of consciousness, and it is this transformative level of learning, both at individual and whole society levels, that radical movement towards sustainability requires.
>
> In any crisis situation, people may stay 'stuck' in first order learning: that is keep 'doing more of the same' which is likely to lead to breakdown, or alternatively, achieve breakthrough which is dependent on reflective, intentional learning which gives rise to new perspectives.
>
> The key point is this: the crisis/opportunity of sustainability requires second – and where possible – third order learning responses by cultural and educational systems. There is a double learning process at issue here: cultural and educational systems need to engage in deep change in order to facilitate deep change – that is, need to transform in order to be transformative. (Sterling & Orr, 2001, p. 9).

I have chosen to present this long citation as I don't think I could say it better myself. It is exactly what Caroline Baillie and Ginie Servant were cited for in Chap.

A. Kolmos, *Engineering Education for Interdisciplinary Futures*, Emergent Directions in Engineering Education 2,
https://doi.org/10.1007/978-3-032-32540-2_9

Table 9.1 Strategies for educational transformation (Kolmos, 2021)

	Add-on strategy	Integration strategy	Re-building strategy
Curriculum changes	Small changes to existing curriculum structure, e.g. new electives	Competency integration into existing courses and high degree of coordination	Crossing or merging disciplines with a high degree of coordination
Educational leadership	Protection and response to external requirements	Strategic coordinator and motivator	Vision, motivator and change agent
Academic staff developments and qualification level	Focus on the individual academic staff level aiming to produce an effective teacher	Focus on the individual and institutional level aiming to produce a skilled and collegial teacher	Focus on the individual, institutional level beyond existing disciplines and aiming to produce a scholarly teacher
Promotion criteria	Visible career tracks no matter which strategy is applied for curriculum change		

2. Sterling and Orr use first, second, and third order learning to conceptualize different stages of change and transformations: the add-on, integration, and rebuilding strategies. We used this framework to identify differences in the curriculum changes and academic development strategies (Kolmos et al., 2016). I developed this framework further to point out variations in educational leadership, academic staff development, and promotion criteria in my article in the second UNESCO Engineering Report (see Table 9.1) (Kolmos, 2021).

Sterling and Orr's categories make sense in systemic change as first and second order changes are within the prevailing societal framework, and third order change is a transformation of current structures, cultures, and educational practices. All strategies are systemic. The add-on strategy is often seen as a heterogeneous strategy, as it maintains existing practice. But indeed, it is also a systemic strategy to keep the existing system and only do smaller adjustments. The integration strategy aims to merge elements in the curriculum, e.g. competencies, interdisciplinary learning, and digitalization, and this strategy needs a strategic coordinator in the organization to facilitate these changes.

However, a rebuilding strategy is about rethinking disciplinary and organizational boundaries. This can include allowing societal challenges to shape the organization's logic, such as recognizing that energy encompasses not only electricity but also all of the disciplines required to contribute to the resolution of societal energy problems.

As previously stated, academic staff development units play a role, and in relation to the three levels of change/transformation, there will be various aims and focal points ranging from individual to collective change agents facilitating and creating visions and common strategies. It does not need to be an entire institution but a department or faculty within an institution, such as University College London (Mitchell et al., 2021).

9.2 Did we Change or Did we Transform?

After 41 years as an active academic working on change agendas for engineering education, I have two relevant questions: Did we change or transform anything at all? Was all the work, and all the higher prioritizing of work and consequently lower prioritizing of family worth it? I think everyone in their late sixties can reflect on this question.

As already shown in the previous chapters, change has happened that concerns both the pedagogical approaches and the inclusion of competencies. During the 1990s and 2000s, there has been a requirement from the political systems to develop higher education in general and engineering education in particular to address societal needs. Employability became urgent; the graduates had to understand what it takes to participate in a work environment to ease the transition from education to work. Part of that was the skills and competency agenda, which was especially necessary for continental Europe and geared towards a much more theoretical approach in the engineering curriculum.

As mentioned before, the competency agenda was aligned to problem-based project work as the students learned to manage projects, share work, and collaborate by integrating knowledge into an innovative process. This was a basic requirement for companies, as technological development is system-based and requires collaborative teamwork.

During my academic career, we changed elements of engineering education to address contemporary societal needs. I was involved in institutional change processes and ran hundreds of workshops over much of the world. Did this, however, make a difference? Did we change education within the existing society's economic and political logic? Did we bring value to future society by educating engineers who are able to respond to climate change and contribute to a greener world? The question here is what is the transformation and what is the change?

If you work within the boundaries of a system, you do not see the need for change. When you start to work on the boundaries of a system, you begin to question the form and construct of the boundaries. Working on the boundaries is a meta-level, where the definition of the boundaries is questionable.

The curriculum changes I have been involved in have been within the self-understanding of engineering education's own boundaries. The CDIO initiative was basically a question of making engineering education more up-to-date and efficient in relation to the transition from education to work. The PBL approach is similar in transforming courses or curricula to become more adequate. But this has not questioned the disciplinary boundaries. To work on the boundaries involves a discussion on content, context, values and directions. This is not to say that the content should be less technical or scientific, but that the context and purpose are reconstructed.

In trying to answer the question 'Did we change anything?', yes, we changed a lot. However, it was within the boundaries and with a purpose of becoming more efficient and relevant for the existing society, which could be shortsighted goals. One can ask if this would not always be the goal for higher education; certainly, but it is a reactive response more than a proactive response.

The points I want to convey here are the reactive/proactive and the short–/long-sighted views. Universities have the responsibility to educate for both short and long-sighted perspectives as well as to educate for both reactive and proactive assessments of technological innovations. Based on science, we need to look forward as well as backward, and we are the institution in society that should be able to make these analyzes on neutral knowledge ground.

This is our obligation, not at least to educate the future generations to take responsibility for making sustainable decisions. Even though the short-sighted goals ought to include development and responsibilities for climate change and equality, the reality is that these goals are long-sighted and all-encompassing in order to educate engineering graduates to be able to participate in the transformation to greener and globally more equal societies as indicated by UN global goals (UNESCO, 2017, 2021).

Poutiatine has written an interesting overview of transformational leadership. He has formulated nine theses for transformational leadership of which the first three are aligned with the previous statements on transformation: that it is not the same as change, requires confirmation to change, and that it is a second-order change (Poutiatine, 2009):

1. Transformation is not synonymous with change
2. Transformation requires assent to change
3. Transformation always requires second-order change
4. Transformation always involves all aspects of an individual's or organization's life
5. Transformational change is irreversible
6. Transformational change involves a letting go of the myth of control
7. Transformational change always involves some aspect of risk, fear, and loss
8. Transformational change always involves a broadening of the scope of worldview
9. Transformation is always a movement towards a greater integrity of identity—a movement toward wholeness. (Poutiatine, 2009, p. 192)

My point in referring to Poutiatine is that transformation is irreversible and will affect all aspects of the individual academic staff or the organization, though I believe it is not an 'or' but an 'and'. Transformation will entail the formation of an identity and an understanding of the surrounding world. Transformation in education will have an impact on both the system and the individual. The process will involve risk, fear, and the letting go of control, which may be one of the most difficult aspects of a transformation process because we tend to rely on our previous experiences and on what is familiar. If something goes wrong during the process, we tend to seek safety, and we did not transform the systems.

9.3 Educational Change at Different Levels

During my career, I have written a great deal on change and transformation. I have been one of the global change agents for engineering education both in terms of describing alternatives to the existing engineering education at the time, and in terms of organizing the field of engineering education research and development. In an article I wrote on successful change I formulated seven premises (Kolmos, 2010):

1. Governance and Educational Policy
2. Educational Research and Political Pragmatism
3. Change at both the System Level and the Individual Level
4. Motivation, Leadership and Visions
5. Implementation Strategy
6. Trust in Students' Learning and Re-selection of Content
7. Regional and Global Communities as Drivers

Fifteen years later these points are still valid, both educational change and transformation include very complicated processes and there are many obstacles in the way. There are several levels influencing educational change and transformation.

Government level: UNESCO addresses the policy at the governmental level, which frames the possibilities for change. In an article in the second UNESCO Engineering report, I formulated five areas at the governmental level: the first two as long-sighted goals (interdisciplinarity and sustainability), followed by creating a research base for any change, and finally educational leadership including motivation of academic staff (Kolmos, 2021).

Accreditation is a tool for governments to steer the development of education, and more and more countries have signed up for the Washington Accord and/or are accredited by ABET. Accreditation can be a disadvantage for institutions wanting to move forward to more student-centered learning as it focus on a traditional course approach. However, it can also be a driver and, as seen from the Washington Accord and ABET, both organizations embrace sustainability and professional competencies (Kolmos, 2021).

Institutional level: The curriculum has changed at many institutions. More student-centered learning methods have emerged, such as active learning in the classroom, the flipped classroom, challenge-based learning, and problem- and project-based learning (PBL) (Chen et al., 2020; Doulougeri et al., 2024; Magnell & Högfeldt, 2015; Reidsema et al., 2017). Practice-related learning, such as internships, industry projects, entrepreneurship, and innovation hubs have all increased (Linn et al., 2004), and we see the integration of the learning of professional competencies (Johnson & Ulseth, 2016) (see also Chaps. 3, 4, and 5). We finally begin to see changes that are not only at the course level but at the system level (Graham, 2018a, 2018b), and which characterize trends in a transformation of engineering education.

Curriculum level: This is where it all comes together for both academic staff framing the learning and for the students experiencing it. Over the last 10 years, we can see aims to include the learning of complexity (Table 9.2), where I have combined elements in the curriculum—the types of problems, knowledge, curriculum structure, and teaching and learning methods—with the other dimensions of the Cynefin framework, ranging from simple to chaotic problems (see Fig. 2.1).

Table 9.2 does not represent reality but there are some interlinked concepts to understand and discuss. It also illustrates some of the changes that happen in

Table 9.2 Combining elements in curriculum development with degree of complexity (Kolmos, 2021)

	Type of problem	Knowlede and competences	Curriculum structure	Teaching and learning methods
Simple	Known problem Known solutions e.g. statics	Disciplines	Subjects/courses	Lectures active learning and flipped classroom
Complicated	Known problem Unknown solutions e.g. zero carbon house	Multi disciplinary	Collaboration among several disciplines	Academic problem-based projects across disciplines
Complex	Unknown problems Unknown solutins e.g. energy zero buildings in energy zero cities New Iot, AI, Bio techologies and sustainability challenges	Inter disciplinary	Re-organisation of the curriculum and development of new student-centered and blended learning models	Complex problem analyses and problem based projects across disciplines and together with stakeholders Mega-projects
Chaotic	Disasters beyond complexity	Training in immedicate action by bringing experiences/problems from chaotic situations into the education		

pockets at an institutional level. We see interdisciplinary learning either in elected courses or across the existing programs addressing complicated problems and to a smaller degree complex problems (Chettiparamb, 2007; Hadgraft & Kolmos, 2020).

For many years we have pointed at the integration of complexity in the curriculum; however, the pandemic period reminded us that we need to be able to handle chaotic problems. This will be the same with the climate situation, where we will have longer periods with both rain and drought. The challenge for engineering education is therefore not only to comprehend complexity but also to address chaotic situations. A re-organization of an educational structure needs a combined top-down management initiative combined with motivated academic staff to carry out educational and cultural changes.

9.4 Values and Culture: A Core Motivation

Motivation for change among the academic staff is both the alpha and omega—otherwise change will not happen. Culture is enacted and constructed (Trowler, 1998); you may say that this is basic knowledge, however, we often forget that we can change the culture and that educational change can be the enabler for a cultural change. We should not be afraid to reach out for something beyond the daily practices.

Self-efficacy and beliefs are important for actions and change, and this has been an approach to analyze and comprehend the motivations of the academic staff. Studies have documented that there are fundamental differences between teacher-centered and learning-centered belief orientations (Samuelowicz & Bain, 2001). These are two totally different roles and ways of approaching students. In the first approach you concentrate on the content to be presented, and in the second one you help students' personal conceptual change (Samuelowicz & Bain, 2001).

There might be a long process from belief to action, and it is worth thinking that even smaller actions will further deepen beliefs (Haney et al., 2002). Educational consultant Julia Atkin has written a paper on values, beliefs, learning principles, and practice with many interesting points. One of them is that change in learning principles and practice originates from values and beliefs in how we learn and what our roles are as teachers or facilitators of learning (Atkin, 1996). Becoming aware of one's beliefs and values might be the first step since many people, academic staff included, can act and practice without a deeper reflection on their own beliefs or underlying values.

If you are very much in a content- or teacher-oriented mode, you will think first of the learning objectives and the presentation of them. The learning is embedded but mostly based on one's own learning experiences. If you are more into student-centered learning, you think of outcomes and how students can digest and learn.

In the recent book *Reshaping of Engineering Education*, we drew the attention to the importance of the alignment between the values of academic staff and the intended changes (Habbal et al., 2024). In Fig. 2.2, I pointed at three different modes dominating the academic values of teaching and learning (Jamison et al., 2014):

- the academic mode with a rather traditional and theoretical approach to learning
- market-oriented mode with a closer collaboration with companies
- the community-oriented mode with an academically critical approach to society and a reform agenda for teaching and learning.

Our point in this book was that most academics do identify themselves with all three modes to some extent. You will not find many academics at universities who deny climate change, the necessity of employability, or the importance of theoretical knowledge. These modes are not contradictory; they represent more ideal types.

You will view teaching theoretical knowledge as the main objective and consider scientific thinking to be fundamental if you are mostly of an academic mindset. You place a far higher emphasis on the utility and application of knowledge if you're more interested in teaching and learning relevant knowledge for future engineering practice, which characterize a market-oriented mindset. A community mindset includes a wider challenge-based perspective, where the analysis and comprehension of larger challenges of both complexity and chaos are feasible.

Change and transformation might start by identifying the mindsets and motivations among the academic staff. They will be the drivers, and they need to believe in what they are expected to do.

9.5 The Power of Cases

Dara Melnyk has worked on success factors for innovative universities that point at personal drive, resources, structural flexibility, and university autonomy. For the dissemination of ideas, it should be seen as a response to contemporary problems and there should be a case that gives some kind of inspiration (Melnyk, 2025).

Fifteen to twenty years ago, it was possible to mention the cases of new student-centered learning systems in engineering education that were known internationally, such as the P5BL model at the following institutions: Stanford University (Fruchter & Lewis, 2003), Olin College (Sommerville et al., 2005), Monterey Tech (Lopez-Islas, 2001), the University of Queensland (Crosthwaite et al., 2006), Central Queensland University (Jorgensen & Howard, 2005), Victoria University (Ozansoy & Stojcevski, 2009), University of Technology and University Tun Hussein Onn (Yusof, et al., 2005), and Singapore Polytechnique (O'Grady & Alvis, 2002). In Europe, these institutions include Aveiro University, Portugal (de Graaff & Kolmos, 2007), University of Mondragon, as well as the universities of Coventry, Newcastle, Sheffield, and Manchester Universities (Graham, 2010). These were the examples mentioned in a chapter in the *Cambridge Handbook of Engineering Education Research* published 2014 (De Graaff & Kolmos, 2014). Since 2014, hundreds of institutions or educational programs have applied principles of student-centered learning.

Cases are powerful. During the decade from 2010, two of my Indian colleagues, Professor Inamdar and Professor Vedula, had formed two very different strategies for trying to establish a change agenda in India. Professor Inamdar established a new college, Vishwaniketan, with a strong inspiration from Aalborg University and mediated by one of my former PhD students Vikas Shinde, to show other Indian institutions that it was possible. A totally different strategy was practiced by Professor Krishna Vedula, who is the founder of the Indo Universal Collaboration for Engineering Education (IUCEE). Professor Vedula invited many American colleagues to India to run courses, with the strategy to get more information out to very many other colleagues.

These two strategies complement each other—you need to know and have information as the first level, but the cases with real practices are convincing as it shows that change is feasible. There are not many who can imagine alternatives and establish totally new practices. Most of us need examples in order to be convinced that it works and be inspired that it can be done a different way.

The learning from cases is multifold. A case is a whole with several variables interacting in processes and products. You can learn from peers how the curriculum can be organized in a different way; imagine what you could do with your own course or program. There is learning from experiencing a different culture and seeing how colleagues argue and talk about their teaching and students' learning. There is also learning from how the change process has been organized and what kind of obstacles have emerged during the process.

There is research on how we learn from comparing cases with special emphasis on similarities and differences (Alfieri et al., 2013). There needs to be similarities to identify that this is within our disciplinary, cultural, or pedagogical identity, and differences indicating the directions for alternatives. The similarities and differences do not need to relate to one variable. If you share culture or knowledge, this will be an entrance to the understanding of the case.

Ference Marton writes about sameness and difference in transfer on the basis of variation theory (Marton, 2006). Variation is a basic principle in all development, e.g. in biology (genetic development) or in learning. Variation in learning is based on the assumption that individuals understand and reflect the world from their individual perspective, and learning takes place when students are "capable of being simultaneously and focally aware of other aspects or more aspects of a phenomenon" (Marton & Booth, 1997, p. 142).

Basically, it is in the same vein as what Vygotsky calls the zone of proximal development, which emphasizes that the intended learning has to be in reach of the learners' experiences and not too far away (Vygotsky, 1978). The advantage of cases is that there are multiple entrances and more possibilities for the academic staff to connect or be within the proximal area for development.

9.6 Systemic Change

Although we find changes in the curriculum, the systemic changes at institutional levels are limited. University College London is a rare example of a re-organization (Mitchell et al., 2021); at MIT institutional change has been made primarily by creating elective threads, and we find the same at Chalmers University with their track program (Bingerud et al., 2023; Lavi et al., 2021). It is fantastic to be able to report on examples where we can see there is a change at a system level. This can be either on the institutional, departmental or most often on the program level (Kolmos, 2017). Most change actually happens by a bottom-up strategy where a single academic can experiment with teaching and learning methodologies as they please.

I have made the case in numerous writings that a systemic PBL approach is crucial to the process of educational reform for both students and academic staff (de Graaff & Kolmos, 2007; De Graaff & Kolmos, 2014; Guerra et al., 2017; Kolmos, 2017; Kolmos et al., 2021). If not, students will encounter island rulers in a curriculum where academic staff control their own courses. In one course, they attend lectures and the next one they meet variations of active learning, and it is left to the students to create coherence both for the content and the way of learning. How the different subdisciplines are linked will often be left to the students.

In many cases, in a final capstone project, students make links to the subdisciplines or to an authentic problem at a company (Dutson et al., 1997; Hauhart & Grahe, 2014; Miller & Olds, 1994). Because capstone projects are completed at the end of an academic program, they also help students make the transition from school to the workplace. Although capstone projects are offered to students quite late in their studies, they might provide an overview of the coherence in their disciplinary learning.

When you have ownership of a single course, you choose the teaching and learning strategies, and if you are one of very few enthusiastic teachers trying out new methods, the institutional and peer support might not be in place. Without a bottom-up approach to change, there will be no cultural change in the learning environment. That should be the first learning for the top management, that if they want to have a more comprehensive change of education, they will need to have change agents at all levels in the organization.

In the many training sessions, I have run over the years, I have listened to participants' experiences of running experiments in single courses. When I first met colleagues trying to apply principles of PBL, they were engaged and energetic—but meeting them again later they were frustrated with not having an overview of the entire curriculum and only making changes within their own course.

What does systemic change actually mean? A systemic approach covers relations among the variables in political, economic, cultural, institutional, and curricular subsystems. However, this will be too comprehensive for institutions to deal with, and those that cannot change the political or economic system will have to maneuver within political frames, including accreditation. For a systemic transformation to PBL at the institutional level, we identified four main factors (Kolmos et al., 2021):

1. Institutional visions and values in relation to different knowledge and problem modes
2. Combination of active teaching and learning methods to facilitate learning of knowledge and generic competencies
3. Interlinked full scale curriculum where the various components are logically related in a progression
4. Explicit outcomes for professional competencies.

The first factor on institutional values involves visions and values. Knoster, Kotter and Fullan mention visions as one of the first basic steps in any change process (Fullan, 2007a, 2007b; Knoster et al., 2000; Kotter, 1995). Without directions for a change process, it will become diffuse and random. Regardless of whether this is a change or a more long-sighted transformation process, an understanding of how individual efforts contribute to the overall picture is important.

Visions and values in the three different university modes are founded in different sets of values, not necessarily contradicting sets of values, but the modes must be seen as complementary (see Fig. 2.2). In a PBL system, the understanding of problem variation is important as the problem leads to the choice of content. If you are working towards a community-based mode, the more complex and comprehensive challenge will determine what academic knowledge you need to learn combined with competencies in a market-oriented mode. If you work from an academic mode, the disciplines will determine the choice of problems, and the application in society and in innovation will be an add-on.

The second factor is the combination of various active learning methodologies. A disadvantage with the Aalborg PBL model is that it is too uniform. Students do a project of 15 ECTS every semester and they learn how to start and conclude the project and how to share work among themselves. Furthermore, they could stay within more or less the same team constellation for several semesters. If you have unified systems, the students become very efficient at learning and acting within these frames. The challenge is to have some kind of variation from lecture, active learning in the classroom, smaller and bigger projects both in terms of credits and durations, and variations in types of problems and projects.

The third factor is coherence and interrelations in the curriculum. (Graham, 2025) has been one of the facilitators of change with reports on project approaches, cases of change processes, and academic development (Graham, 2010, 2012b; Graham, 2017; Graham, 2018a, 2022a, 2022b). Her reports are well recognized academically and have had an impact on the development of engineering education worldwide. In one of her early reports, she emphasizes coherence in the curriculum and the fact that the modules are interlinked and not isolated islands. This is aligned with the criticism of a curriculum where students are expected to create the coherence.

The fourth factor embraces the explicit learning outcomes and assessment for professional competencies. As discussed in the previous section, the learning of professional competencies should not be tacit or be just an added value. To learn teamwork skills and competencies, there is the need for experiments (variation) and reflections (see Chap. 6).

The CDIO approach I referred to in Chap. 5 has been an ongoing network for engineering education, embracing a system approach to the curriculum (Edström & Kolmos, 2014). What characterizes CDIO to many other networks is that it has a focus on systemic change as well as change at the single course level. Three factors are present in the CDIO system approach, but the first one on visions and values is not really visible.

In a systemic change process I would add a fifth criterion—that academic staff at all levels need to be involved. That means that we need both a top-down and a bottom-up approach (De Graaf & Kolmos, 2007). Developing new cultures at all levels in an organization is important. The best situation will be that academic staff have experimented with teaching and learning methodologies for a while, after which there will be experiences and motivations to build upon. This should be one of the first initiatives that top management initiate, as a slowly emerging motivation for change at a system level will be essential for success.

9.7 Alignment and Misalignment in the Curriculum as Drivers

Alignment among the elements of the curriculum will enhance students' learning. Biggs indicated that there should be interrelations between the three Ps: the 'Presage' with students' prior knowledge and the teaching content; the 'Process' by learner-focused activities with surface and deep learning; and the 'Product' with the learning outcomes (Biggs, 2003; Biggs & Tang, 2011). In Scandinavia there is a relational curriculum model with selected elements: goals, content, learning, assessment, and teachers and the relations among these elements (Hiim & Hippe, 2001).

The understanding of the curriculum as an interrelation among the curriculum elements was new compared to earlier approaches in Scandinavia. During the 1960s, the approach was mostly a planning tool and later evolved into an analytical model for planning and implementing teaching and learning, which might be more simplistic as the most dominating teaching methods were lectures and, to a small degree, laboratory work (Kolmos, 2002; Laursen, 1998). As more student-centered learning methodologies gained ground, there was a need for models with more elements to analyze, understand, and design learning processes with students driving the learning process. There was a move from 'just' organizing how to present the content to how to organize the students' learning process of the content.

Erik de Graaff and I applied the relational curriculum model for developing a PBL curriculum model to indicate the importance of a more holistic change. We identified seven elements in the PBL curriculum which should be aligned including objectives, problems, progression, students' learning, academic staff, space and organization and assessment. If the learning outcomes are to identify and analyze the variables in sustainable traffic in cities, then it will not make sense to give the students a specific task on traffic regulation or conduct multiple choice tests as assessment method (Fernández-Samacá et al., 2011; De Graaff & Kolmos, 2014).

In principle, the idea with both Biggs's P-model and the relational curriculum model is that there should be an alignment in order to achieve success in a curriculum. All the elements have to support each other, and especially the assessment system must support the intended learning outcomes, otherwise the students will draw their attention to the assessment criteria instead of the learning outcomes.

Assessment was mentioned as a driver for change, and if the assessment was not changed accordingly with the learning outcomes, it would create a barrier for learning. On the other hand, a change in assessment and being explicit about the criteria could be a driver for enhancing students' learning (Gibbs, 1999, 2006).

At the system level, an appropriate alignment would be preferable. If there is too much misalignment, it would be difficult to achieve the intended learning outcomes. As described in Chap. 4, this was one of our concerns when the Danish Ministry for Higher Education in 2006 prohibited group assessment for the project teams. At that time, we were concerned that the PBL system would not be able to survive.

This story on alignment is not so much about group assessment, but as an example of a misaligned system that worked despite the conditions. The formal and informal curriculum, or the explicit and hidden curriculum, exist side by side, and it is not only the formal and explicit curriculum that dominates the development. In this case, it was the informal curriculum and the power of academics' values, beliefs, and cultural practices that maintained and developed the PBL system.

However, the informal curriculum, dominated by existing culture, might provide resistance for a long time by finding ways around it. In this case it was to keep the PBL system but in other change processes it might be to keep another prevailing system. Curriculum changes at the formal level might be rather easy to accomplish, but the informal curriculum takes time. I will return to this in Chap. 10 on educational development units.

9.8 Management and Leadership in Successful Change

In the book *Management of Change,* we only talked about management without a clear outline of the leadership dimension (de Graaff & Kolmos, 2007). Uljens and Ylimaki point out that while leadership has been viewed much more as an administrative and management role in Europe, it is a different subject in some other parts of the world, such as the US, where it is paired with a governance perspective. As a result, in a European setting the leadership position has been less prestigious and developed (Uljens & Ylimaki, 2017).

Ruth Graham has called for the development of educational leadership and emphasized the need of educating leaders. The development of educational leadership includes both the training of leaders and the creation of an organizational structure that facilitate the impact of leadership (Graham, 2012a, 2022b). Educational leadership must be a synthesis of organizational and curriculum theories so that the president, deans, heads of departments and schools, and others all have distinct responsibilities to play and contribute to the directions of higher education. Maybe we approach distributed leadership at different levels in the organization here. A common vision, a shared grasp of the implementation procedures, and an understanding of how to facilitate these processes are all necessary to bringing about transformation or change among the layers of leaders.

There is not so much about leadership in the literature on engineering education. One of my latest PhD students, Henrik Worm Routhe, wrote about student leadership in interdisciplinary system projects (Routhe et al., 2023; Routhe et al., 2024a, 2024b). In his work there is a focus on design processes and what to take into consideration when designing interdisciplinary projects, such as vision, educational methodologies, and support structures (Klaassen, 2018; Van den Beemt et al., 2020).

Most of the design and curriculum recommendations go in the direction of team-based project work to create opportunities for students to learn interdisciplinary collaborative competencies. However, even if studies do focus on the design and planning process, fewer studies include the actor perspective, such as the roles of the leaders, coordinators, and facilitators in the transformation processes of an interdisciplinary curriculum. Several Finnish studies delve into the perception of interdisciplinarity among the engineering faculty and this varies according to the discipline (Feng & Hölttä-Otto, 2021, 2022; Feng et al., 2023).

Successful leaders who managed to carry out a successful change have tried many times before with less successful changes; they have experiences in their backpack to build upon. Failures are not very often reported in the literature, but they do exist, and without the experiences of how not to do things there will be fewer successes. Experiences from educational leadership and management are important elements in any change process.

A review of educational leadership in higher education highlights the important aspects of leadership. These qualities and acknowledgments, and respect are not only something you can learn, but are earned during academically solid and respectful work (Bryman, 2007, 2013; Habbal et al., 2024).

Table 9.3 provides an overview of the various competencies that are affiliated with an educational leader. Most of my research and writings during the years have focused on the function more than on the person. However, it's never too late to learn, and I have slowly admitted that the person and their qualifications in terms of skills, competencies, and traits are important. Sometimes a person—a charismatic leader—can move an organization despite the structural barriers, but this is in rare situations. Even charismatic leaders will be restricted by structural boundaries and, whereas Bryman (2007) approaches leadership at a personal level, we need to see leadership as an organizational function.

Educational leadership is not the same as educational management. Leadership embraces visions, values, directions, political influence, external partnerships, strategies, employee motivation, and the creation of a constructive working environment (Bush, 2003, 2007). Management concerns the daily routines, plans, economy, time

Table 9.3 Competencies in educational leadership thematized (Habbal et al., 2024)

Vision and direction	• Clear sense of direction/strategic vision • Preparing departmental arrangements to facilitate the direction set
Integrity and respect	• Being considerate • Treating academic staff fairly and with integrity • Being trustworthy and having personal integrity • Acting as a role model/having credibility
Influence and communication	• Allowing the opportunity to participate in key decisions/encouraging open communication. • Communicating well about the direction in which the department is going. • Creating a positive/collegial work atmosphere in the department.
Recognition	• Advancing the department's cause with respect to constituencies internal and external to the university and being proactive in doing so. • Providing resources for and adjusting workloads to stimulate scholarship and research. • Making academic appointments that enhance the department's reputation.

schedules, and responsibilities for staff development. Leadership and management are two different functions in educational development and change. Leadership is affiliated with long term plans, and management with the short-term plans as action plans with activities, resources and time. Part of the action plan can include the education of change agents, training of academic staff, incentives and resources for change (Knoster et al., 2000; Kotter, 1995).

Lippitt and Knoster (Fig. 9.1) made a famous scheme for the necessity of all elements and what happens if one of the elements is missing (Knoster et al., 2000; Lippitt, 1987). This is a simplistic version of leadership and management in a change process; however, as simple messages underpin the meanings, it illustrates the importance of a holistic and systemic view on the educational change in an organization.

Visions, consensus in the organization, skills, and incentives are all elements in the leadership function. All elements are necessary in a change process and if one of the elements is missing, it might lead to confusion, sabotage, anxiety, resistance, frustration or a treadmill. This is a very straightforward overview of some of the elements in a change process.

Leadership and management culture has changed over the last 30 years. The trend has been to move from elected leaders to appointed leaders (in Europe), which indicates a general move from a bottom-up approach to a top-down approach in management. With that it might be easier to make decisions but not necessarily to carry out changes.

For the development of vision in an institutional change process it is important to involve academic staff in that formulation, as it creates ownership and motivation. At the same time in Denmark, we experience increased governance on the regulation of seats, degrees, and resources, to which institutions must respond. In a political scene like this it can be difficult to create reliable visions at the institutional level.

Vision +	**Consensus +**	**Skills +**	**Incentives +**	**Resources +**	**Action Plan**	**= Change**
	Consensus +	Skills +	Incentives +	Resources +	Action Plan	**= Confusion**
Vision +		Skills +	Incentives +	Resources +	Action Plan	**= Sabotage**
Vision +	Consensus +		Incentives +	Resources +	Action Plan	**= Anxiety**
Vision +	Consensus +	Skills +		Resources +	Action Plan	**= Resistance**
Vision +	Consensus +	Skills +	Incentives +		Action Plan	**= Frustration**
Vision +	Consensus +	Skills +	Incentives +	Resources +		**= Treadmill**

Fig. 9.1 Lippett and Knoster's overview of elements in a change process (Knoster et al., 2000; Lippitt, 1987)

With any change, there is a need to educate change agents, and their role is to be the frontiers of change and to give inspiration to others. Change agents can be affiliated with many levels in the organization and they can be formal leaders or engaged colleagues who have already tried out several student-centered experiments.

Top management must initiate the education of change agents, and generally, they should not be the ones to carry through the changes but the ones to frame the change or transformation process. University leaders are rarely trained as leaders. Due to the management load, they lose control and interaction with the teachers, and no matter the reason, management will distance itself from the teachers. Distance will then come and along with that suspicion and lack of trust. Distance creates mistrust, and new initiatives from the top might be perceived with mistrust. Academic staff begin to react with strategies to avoid influence on existing practice and this is not an atmosphere for a change culture.

Educational leadership also includes the promotion systems of universities. These promotion systems focus predominantly on research by the number of journal articles, impact factors, and the amount of external funding. Teaching qualifications come second and even if it says that they should count, there is a widespread anticipation and culture that you need to focus on your research. Ruth Graham not only emphasized the need for educational leadership but she developed a qualification framework for teaching qualifications (Graham, 2015).

A promotion system will acknowledge these different levels at an explicit level. The explicitness is an important point, as teaching qualifications in most places are more or less implicit, and sometimes a teacher is awarded 'teacher of the year' without explicit criteria. When you have visible and explicit criteria in a system, it will drive the employees' orientation and values. To create this is part of a leadership function and to get this out into a cultural practice is a leadership success.

9.9 Stagnation, Institutionalism, and Trust

> *To sustain the implementation of new ideas, the key factors are personal drive and institutional support in the form of resources; structural flexibility; positive perception; and university autonomy* (Melnyk, 2025, p. 8).

What Dara Melnyk indicates is that a change or a transformation is not forever; it needs to be cultured, energized, and developed. When institutionalization has really taken hold, there is always a danger that change will stagnate. Stagnation is not a positive concept as it reminds one of routines and a kind of instrumentalism. It is when curiosity dies and there is no movement, but there is practice as usual.

We talk about stagnation in the economy, politics, institutions, and innovation; however, very seldom do we see it used as a concept in the literature on engineering education or on higher education in general. If we are to understand the change and transformation processes, we also need to examine the phases where everything stops or as a wave goes back and forth.

Sometimes there is a change in the leadership and a choice of rolling back the changes in the curriculum. History has shown this several times in engineering education, and one of the cases I was involved in during the early 2000s was Leuven-la-Neuve in Belgium (Frenay et al., 2007). A change of leadership can result in an unchanged curriculum.

The distinction between educational management and leadership is also embedded in the understanding of stagnation. As already described in this section, educational management and leadership are not the same. In rough terms, leadership is about pointing out the direction, whereas management is about the daily routines. If you only do management, the system will stagnate; if you only do leadership, the organization will be a mess. It is about finding a balance even if a change has been institutionalized to keep the visions alive and develop them further. As Fullan describes, there will be a phase of institutionalization after a change where the new system needs to find a balance and come into an operational mode (Fullan, 2007a, 2007b).

Institutionalization is a must in any change process, as the system needs to work and be manageable. The risk, however, is that it becomes a new routine and the visions and change efforts somehow just come to a standstill. In and across courses, academics fall back into the newly established educational practices and stop being curious of future possibilities. For example, this can be seen whilst presenting the students with the same catalogue of problems or challenges to work on, or with the same procedures for forming the project teams or ending up with the same number of students in the project teams.

The motivation for experimenting and initiating variation in the project facilitation practices vanishes in the same way as for a lecturer who has been using the same slides for several years. In the interdisciplinary projects, e.g., continuing the electric car project for further optimization would not be as stimulating and motivating as the first interdisciplinary team to construct the foundation and the visions for the car would experience.

One could expect that in a problem- and project-based curriculum, the emphasis on the problem would secure innovation and continuous improvements. However, as shown in Chaps. 3 and 4, the character of the problem can vary, ranging from more narrow disciplinary problems to broader challenges. If students work with the broader and more open problems, where the first phase identifies the real problem, this then influences their learning process and the way the facilitator interacts with them.

For academics, routines are necessary for surviving the ever-decreasing resources for teaching and pressure for publications. Although, routines are also demotivating for both academics and students as there is no discovery, curiosity, or drive to try and establish new, unknown learning experiences. For students this might work, as they are aware that they are to gain a certificate and learn content, but it might not motivate them to walk the extra mile. Orientation to problems and challenges involves openness and curiosity. This is the recipe to keep going in an active learning curriculum, especially a PBL one, and a routine approach is an issue for longer periods.

At an institutional level, there needs to be periods of consolidation before further change can happen, and before moving on experiences need to be summed up and learning from new practices evaluated. There is also a need to cultivate visions and trial new practices to meet these visions. A PBL curriculum does not secure innovation, creativity, or deep learning; it depends on the interaction between the culture and the formal curriculum.

Olin College was founded with no tenure track system nor with faculty or institutionalized boundaries. The idea was to avoid scientific staff becoming too accustomed to routines and yesterday's practices (Sanoff, 2000). The experiments with new challenges across new courses are important to maintain and facilitate students' curiosity and engagement (Miller, 2010; Miller & Euchner, 2014). Several studies have emphasized that at the end of the day what means most to the students is the academic staffs' engagement in the subject.

Avoiding faculty boundaries and tenure tracks might be radical; however, a very important condition is that there is organizational trust. This is possible at smaller places where a new practice is created, the change is not caused by economic reasons, and where you know most of your colleagues or at least some in the other clusters of academic staff.

Trust is a must in an organization on the move. Institutions that have changed to a partial project approach during a phase of economic reductions will have mistrust among the academics. If there is a suspicion of staff reductions, fear and anxiety will dominate the atmosphere. If there is trust between top and bottom, change processes are much easier to carry out, as motivation and desire to experiment have possibilities for growth. In small organizations, trust is easier to build as compared with large organizations.

Too big a system will create distances between the academic staff on the ground and top management. There is no real time to reach out to employees, and if this happens, it might have more the character of a hearing. A hearing can adjust details but might not influence the basic idea as these already are formed.

In large organizations there will be a distance between top and bottom by nature. Especially in an educational system's perspective, where the entire institution must follow a set of rules for a pedagogical model, it creates even more sincere distances as it concerns the teaching in the classroom. As well as research creating identities, teaching practices create identities and will be experienced partly within the personal sphere.

Trust is the foundation for creative development of pedagogy and culture. It is the foundation for the individual to take responsibility. It requires spaciousness, perspective, and vision. Fear is the foundation for stagnation and defense, often caused by one-way communication, mass communication, orders, and restrictions. The trust and loyalty that has existed between the bottom and the top will disappear.

With distance, a tougher management style will also emerge; management no longer facilitates, they decide. A hierarchy develops and with distance comes fear and anxiety, which is not good for anyone in an organization. If you are anxious, you say nothing but go under the radar. An organization with people under the radar is an incoherent organization—there is no playing on the field from the bottom but rather 'obeying'. It basically becomes an unhealthy organization in terms of further development if the decisions are made at the top and the employees feel they just implement the ideas.

This might be a danger for any institution, including the PBL institutions such as Aalborg University. Having PBL-principles as a strategic profile this is a great idea, and as mentioned in the first chapters it is partly because of being famous around the world (Graham, 2018b). However, with the institution growing larger this might now have the difficulty in creating a common identity and there might be a need for more decentralized pedagogical visions, cultures and identities.

There is a limit in the size of an organization for employees to feel ownership and control of their own situation. If it becomes too big, not only should the economy be decentralized, but also the pedagogical ideas should be owned by the academic staff. There can be an interaction with the central level, but the feeling of integrity in the teaching and learning is essential for the energy that will be given in the process. Nothing is ever simple or black and white. But perhaps for top-driven institutions, it is time for the pendulum to swing again.

9.10 Reflection 9: Living-With Chaos

Both the top and bottom of an organization must address the challenge of living-with chaos. Leaders and management must begin to take this seriously; not only is there an increasing number of young people with diagnoses, but there is also a general sense of anxiety in society.

If leaders begin to talk, then academic staff may follow. Articulation is the first step and this is what Richard Miller, the former president of Olin College, is doing in the establishment of the Coalition for Transformative Education, renamed

LearningWell (Miller, 2025). It's about caring for the students, not in the sense of nursing, but to be 'seen'.

Students need to be met in education, viewed as people and personalities and not as numbers in a general mass. This is hard given the current conditions for higher education and could put extra pressure on the teachers. In general, we hear about the cuts of resources in higher education and the difficult times for both research and teaching. Much of the employees' time must be spent on finding money for the next research project, and attention is taken away from students.

We also need a wake-up call for the leaders in engineering education. The universities have a huge responsibility in educating the next generation of engineers, and there is a need to focus on the whole person not just the knowledge and competency parts. The leaders will set the agenda not only by words but with their actions and personal integrity. Integrating living-with chaos in education will require leadership and first and foremost a vision for students' learning.

There have been several initiatives to move to a more mission-driven university which should indeed be an interdisciplinary initiative. Maybe this could be combined with boundary crossing, resilience and moving to empowerment of students as well. Students need to be in the core of the strategies.

References

Alfieri, L., Nokes-Malach, T. J., & Schunn, C. D. (2013). Learning through case comparisons: A meta-analytic review. *Educational Psychologist, 48*(2), 87–113.

Atkin, J. (1996). *From values and beliefs about learning to principles and practice*. IARTV.

Biggs, J. (2003). *Teaching for quality learning at university* (2nd ed.). Open University Press.

Biggs, J., & Tang, C. (2011). *Teaching for quality learning at university* (4th ed.). Open University Press.

Bingerud, M., Enelund, M., Briggs, K. H., Karlsson-Bengtsson, A., & Larsson, J. (2023). Tracks: Impactful reform for flexible adaptable education. *Transforming Engineering Education, 23*.

Bryman, A. (2007). Effective leadership in higher education: A literature review. *Studies in Higher Education, 32*(6), 693–710.

Bryman, A. (2013). *Leadership and organizations (RLE: Organizations)*. Routledge.

Bush, T. (2003). *Theories of educational leadership and management*. Sage.

Bush, T. (2007). Educational leadership and management: Theory, policy and practice. *South African Journal of Education, 27*(3), 391–406.

Chen, J., Kolmos, A., & Du, X. (2020). Forms of implementation and challenges of PBL in engineering education: A review of literature. *European Journal of Engineering Education*, 1–26.

Chettiparamb, A. (2007). *Interdisciplinarity: A literature review*. HEA Interdisciplinary Teaching and Learning Group, Centre for Languages, Linguistics and Area Studies, University of Southampton.

Crosthwaite, C., Cameron, I., Lant, P., & Litster, J. (2006). Balancing curriculum processes and content in a project centred curriculum – In pursuit of graduate attributes. In Education for Chemical Engineers. Trans IChemE, Part D, no. 1, pp. 39–48. Retrieved from http://www.icheme.org/ECEsamplepaper.pdf

de Graaff, E., & Kolmos, A. (Eds.). (2007). *Management of Change: Implementation of problem-based and project-based learning in engineering*. Sense Publishers.

Doulougeri, K., Vermunt, J. D., Bombaerts, G., & Bots, M. (2024). Challenge-based learning implementation in engineering education: A systematic literature review. *Journal of Engineering Education., 113*, 1076–1106.

Dutson, A. J., Todd, R. H., Magleby, S. P., & Sorensen, C. D. (1997). A review of literature on teaching engineering design through project-oriented capstone courses. *Journal of Engineering Education, 86*(1), 17–28.

Edström, K., & Kolmos, A. (2014). PBL and CDIO: Complementary models for engineering education development [article]. *European Journal of Engineering Education, 39*(5), 539–555. https://doi.org/10.1080/03043797.2014.895703

Feng, X., & Hölttä-Otto, K. (2021). *An investigation of the influence of disciplinary distance in interdisciplinary education through Faculty's experience*. International Design Engineering Technical Conferences and Computers and Information in Engineering Conference.

Feng, X., & Hölttä-Otto, K. (2022). *An exploration of teachers' competencies in interdisciplinary engineering education*. International Design Engineering Technical Conferences and Computers and Information in Engineering Conference.

Feng, X., Ylirisku, S., Kähkönen, E., Niemi, H., & Hölttä-Otto, K. (2023). Multidisciplinary education through faculty members' conceptualisations of and experiences in engineering education. *European Journal of Engineering Education*, 1–17.

Fernández-Samacá, L., Nielsen, K. M., Ramírez, J. M., & Kolmos, A. (2011). Comparison of PBL curricula within control engineering education. *International Journal of Engineering Education, 27*(6), 1362–1373.

Freire, P. (1973). *Education for critical consciousness*. Bloomsbury Publishing Kindle edition.

Frenay, M., Galand, B., Milgrom, E., & Raucent, B. (2007). Project-and problem-based learning in the engineering curriculum at the University of Louvain. In *Management of Change* (pp. 93–108). Brill.

Fullan, M. (2007a). *Leading in a culture of change (Revised edition ed.)*. Jossey-Bass.

Fullan, M. (2007b). *The new meaning of educational change, Fourth Edition* (4th ed.). Teachers College Press.

Gibbs, G. (1999). *Using assessment strategically to change the way students*. Assessment matters in higher education, 41.

Gibbs, G. (2006). How assessment frames student learning. In *Innovative assessment in higher education* (p. 23).

Graham, R. (2010). *UK approaches to engineering project-based learning white paper sponsored by the Bernard M*. Gordon-MIT Engineering Leadership Program.

Graham, R. (2012a). *Achieving excellence in engineering education: The ingredients of successful change*.

Graham, R. (2012b). Achieving excellence in engineering education: The ingredients of successful change. *The Royal Academy of Engineering, 3*.

Graham, R. (2015). *Does teaching advance your academic career?: Perspectives of promotion procedures in UK higher education*. Royal Academy of Engineering.

Graham, R. (2017). *Snapshot review of engineering education reform in Chile*. Alborg University/ Pontificia Universidad Católica de Chile.

Graham, R. (2018a). *The career framework for university teaching: Background and overview*. http://www.rhgraham.org/page/rewarding-teaching/.

Graham, R. (2018b). *The global state of the art in engineering education*. M. New Engineering Education Transformation. neet.mit.edu.

Graham, R. (2022a). *CEEDA case study report: Six case studies of the impact of COVID-19 on global practice in engineering education*.

Graham, R. (2022b). *Crisis and catalyst: The impact of COVID-19 on global practice in engineering education*.

Graham, R. (2025). *Teaching cultures*. https://teachingcultures.com/Findings/2025/

Guerra, A., Ulseth, R., & Kolmos, A. (2017). *PBL in engineering education: International perspectives on curriculum change*. Sense Publishers. https://doi.org/10.1007/978-94-6300-905-8

Habbal, F., Kolmos, A., Hadgraft, R. G., Holgaard, J. E., & Reda, K. (2024). *Reshaping engineering education: Addressing complex human challenges*. Springer Nature.

Hadgraft, R. G., & Kolmos, A. (2020). Emerging learning environments in engineering education [article]. *Australasian Journal of Engineering Education, 25*(1), 3–16. https://doi.org/10.1080/22054952.2020.1713522

Haney, J. J., Lumpe, A. T., Czerniak, C. M., & Egan, V. (2002). From beliefs to actions: The beliefs and actions of teachers implementing change. *Journal of Science Teacher Education, 13*(3), 171–187.

Hauhart, R. C., & Grahe, J. E. (2014). *Designing and teaching undergraduate capstone courses*. John Wiley & Sons.

Hiim, H., & Hippe, E. (2001). *Å utdanne profesjonelle yrkesutøvere*. Gyldendal akademisk.

Jamison, A., Kolmos, A., & Holgaard, J. E. (2014). Hybrid learning: An integrative approach to engineering education [article]. *Journal of Engineering Education, 103*(2), 253–273. https://doi.org/10.1002/jee.20041

Jorgensen, D., & Howard, P. (2005). *Ten years in the making – a unique program in engineering*. Paper presented at WACE 2005.

Johnson, B., & Ulseth, R. (2016). *Development of professional competency through professional identity formation in a PBL curriculum*. Frontiers in education conference (FIE), 2016 IEEE,

Klaassen, R. G. (2018). Interdisciplinary education: A case study. *European Journal of Engineering Education, 43*(6), 842–859.

Knoster, T., Villa, R., & Thousand, J. (2000). A framework for thinking about systems change. In *Restructuring for caring and effective education: Piecing the puzzle together*. Paul H. Brookes Publishing.

Kolmos, A. (2002). Facilitating change to a problem-based model. *International Journal for Academic Development, 7*(1), 63–74. https://doi.org/10.1080/13601440210156484

Kolmos, A. (2010). Premises for changing to PBL. *International Journal for the Scholarship of Teaching and Learning, 4*, 1–7.

Kolmos, A. (2017). PBL curriculum strategies: From course based PBL to a systemic PBL approach. In *PBL in engineering education: International perspectives on curriculum change* (pp. 1–12). Sense Publishers. https://doi.org/10.1007/978-94-6300-905-8

Kolmos, A. (2021). Engineering education for the future. In *Engineering for sustainable development* (pp. 121–128). UNESCO.

Kolmos, A., & Graaff, E. d. (2014). Problem-based and project-based learning in engineering education: Merging models. In B. M. Olds & A. Johri (Eds.), Cambridge handbook of engineering education research. (pp. 141–161.). : Cambridge University Press.

Kolmos, A., Hadgraft, R. G., & Holgaard, J. E. (2016). Response strategies for curriculum change in engineering. *International Journal of Technology and Design Education, 26*(3), 391–411. https://doi.org/10.1007/s10798-015-9319-y

Kolmos, A., Holgaard, J. E., & Clausen, N. R. (2021). Progression of student self-assessed learning outcomes in systemic PBL. *European Journal of Engineering Education, 46*(1), 67–89.

Kotter, J. P. (1995). Leading change: Why transformation efforts fail. *Harvard Business Review, 73*, 59–67.

Laursen, P. F. (1998). Didaktikkens kognitionsbegreb. *Psyke & Logos, 19*(2).

Lavi, R., Bathe, M., Hosoi, A., Mitra, A., & Crawley, E. (2021). The NEET ways of thinking: Implementing them at MIT and assessing their efficacy. *Advances in Engineering Education.*

Linn, P. L., Howard, A., & Miller, E. (2004). *Handbook for research in cooperative education and internships*. Routledge.

Lippitt, M. (1987). *The managing complex change model*. Enterprise Management.

Lopez-Islas, J. R. (2001). *Collaborative learning at Monterrey Tech-Virtual University*. Paper presented at the Invited Symposium on Webbased Learning Environments to Support Learning at a Distance: Design and Evaluation. Asilomar, Pacific Grove, CA.

Magnell, M., & Högfeldt, A.-K. (2015). *Guide to challenge driven education*. KTH Royal Institute of Technology.

Marton, F. (2006). Sameness and difference in transfer. *The Journal of the Learning Sciences, 15*(4), 499–535.

Marton, F., & Booth, S. (1997). *Learning and awareness*. Lawrence Erlbaum Associates.

Melnyk, D. (2025). *Success factors of innovative universities*.

Miller, R. (2010). *From the ground up: Rethinking engineering education for the 21st century symposium on engineering and Liberal education*. Union College.

Miller, R. (2025). *LearningWell coalition*. https://learningwell.org/the-coalition-for-transformational-education-announces-new-name-the-learningwell-coalition/
Miller, R., & Euchner, J. (2014). The future of engineering education: An interview with rick miller. *Research-Technology Management, 57*(1), 15–19.
Miller, R. L., & Olds, B. M. (1994). A model curriculum for a capstone course in multidisciplinary engineering design. *Journal of Engineering Education, 83*(4), 311–316.
Mitchell, J. E., Nyamapfene, A., Roach, K., & Tilley, E. (2021). Faculty wide curriculum reform: The integrated engineering programme. *European Journal of Engineering Education, 46*(1), 48–66.
O'Grady, G., & Alvis, W. A. M. (2002). *One day one problem: PBL at Republic Polytechnic*. Paper presented at 4th Asia-Pacific Conference on PBL, Hatyai, Thailand.
Ozansoy, C., & Stojcevski, A. (2009). *Problem-based learning in electrical and electronic engineering education*. In X. Du, E. de Graaff, & A. Kolmos (Eds.), *PBL – Diversity in research questions and methodologies*. Rotterdam: Sense.
Poutiatine, M. I. (2009). What is transformation?: Nine principles toward an understanding of the transformational process for transformational leadership. *Journal of Transformative Education, 7*(3), 189–208.
Reidsema, C., Kavanagh, L., Hadgraft, R., & Smith, N. (2017). *The flipped classroom: Practice and practices in higher education*. Springer.
Routhe, H. W., Holgaard, J. E., & Kolmos, A. (2023). Experienced learning outcomes for interdisciplinary projects in engineering education. *IEEE Transactions on Education*, 1–13. https://doi.org/10.1109/TE.2023.3284835
Routhe, H. W., Holgaard, J. E., & Kolmos, A. (2024a). Leadership in interdisciplinary engineering students' projects: A faculty perspective for supporting the development of student leadership. *International Journal of Engineering Education, 40*(4), 1–19.
Routhe, H. W., Holgaard, J. E., & Kolmos, A. (2024b). Students' learning of management and leadership in engineering education–a literature review. *European Journal of Engineering Education, 49*(3), 540–576.
Samuelowicz, K., & Bain, J. D. (2001). Revisiting academics' beliefs about teaching and learning. *Higher Education, 41*, 299–325.
Sanoff, A. P. (2000). Creating a masterpiece at Olin College. *ASEE Prism, 10*(1), 20.
Sterling, S., & Orr, D. (2001). *Sustainable education: Re-visioning learning and change* (Vol. 6). Green Books for the Schumacher Society.
Trowler, P. R. (1998). *Academics responding to change*. New Higher Education Frameworks and Academic Cultures.
Uljens, M., & Ylimaki, R. M. (2017). *Bridging educational leadership, curriculum theory and Didaktik: Non-affirmative theory of education*. Springer Nature.
UNESCO. (2017). *Sustainable development goals*. http://en.unesco.org/sdgs
UNESCO. (2021). *Engineering for sustainable development: Delivering on the sustainable development goals*. UN.
Van den Beemt, A., MacLeod, M., Van der Veen, J., Van de Ven, A., van Baalen, S., Klaassen, R., & Boon, M. (2020). Interdisciplinary engineering education: A review of vision, teaching, and support. *Journal of Engineering Education, 109*(3), 508–555.
Vygotsky, L. (1978). Interaction between learning and development (M. Cole, Trans.). In *Mind in society* (pp. 79–91). Harvard University Press.
Yusof, K., Tasir, Z., Harun, J., & Helmi, S. A. (2005). Promoting problem-based learning (PBL) in engineering courses at the Universiti Teknologi Malaysia. Global Journal of Engineering Education 9(2) pp. 175-184. Retrieved from http://www.wiete.com.au/journals/GJEE/Publish/vol9no2/Yusof.pdf.

Chapter 10
Educational Development Units and the Power of Networking

10.1 Introduction

As mentioned in Chap. 9, educational leadership can exist at multiple levels in an organization, and it is not about making decisions; rather, it points to visions and directions for decisions and, in the case of clever leadership, the delegation of decision-making to employees within the organization. The latter requires courage and trust, and great leadership is also the facilitation of bottom-up processes by giving employees ownership that can be motivating.

Educational development units and international networks have played critical roles in the spread of student-centered learning approaches. The German saying "*so ein Ding müssen wir auch haben*" illustrates the mechanism of information flow in networks and serves as an inspiration for alternative teaching and learning methods.

I believe that educational development units and the international network contribute to educational leadership both internationally and within a single institution, and they deserve to be recognized. This is not always the case, unfortunately. As it has been difficult to gain respect for PBL, particularly when teaching PBL competencies to engineering students, educational development units share the difficult process of gaining recognition in organizations. Engineering education researchers have been particularly affected by this, as has the establishment of a field that is fundamentally different from engineering research. Engineering education research (EER) is an interdisciplinary field that combines the knowledge of engineering processes with educational research, which falls under humanities and, in some cases, social science. The building of this community during the decade of the 2000s has raised many discussions on profiles and research approaches (Borrego et al., 2008; Buckley et al., 2023; Finelli et al., 2016; Jesiek et al., 2009).

Many EER professionals have undergone a hybridization process, either by transitioning from engineering to educational research or vice versa, which has resulted in a new practice. This has had personal implications for the professional identity in

A. Kolmos, *Engineering Education for Interdisciplinary Futures*, Emergent Directions in Engineering Education 2,
https://doi.org/10.1007/978-3-032-32540-2_10

this area, and several studies on the scope and profile of this new field were conducted in the early years (Brunhaver et al., 2018; London et al., 2021).

It is a role full of dilemmas, and at a personal level it requires the capability to navigate in troubled waters, especially if there is a vision for learning in engineering education and the intention of making a difference.

10.2 Educational, Academic Staff or Faculty Development?

Faculty development, staff development, academic staff development, or educational development are all concepts trying to address the same thing. Again, there are many concepts. Gibbs argues that we should view this as educational development, and that academic staff is essential for advancing education (Gibbs, 2013). I entirely agree with him; however, I have retained the concept of academic staff development because the change to PBL is one of educational development, and academic staff are the core component of that process.

Graham Gibbs is a retired UK researcher in higher education and one of the founders of the International Consortium for Educational Developers (ICED), which is the connecting point for many academic staff development units in Europe. He has outlined activities undertaken by educational developers, ranging from educating the individual teacher to the institution, and the reflections on the development in the approach to academic staff development (Gibbs, 2013). According to Gibbs, the trend in the learning theories and learning regimes moves from teaching content to organizing student learning. Some of these are listed below (Gibbs, 2013):

- From a focus on the classroom to a focus on the learning environment
- From a focus on individual teachers to a focus on course teams, departments, and leadership of teaching
- From a focus on teaching to a focus on learning
- From small, single, separate tactics to large, complex, integrated, aligned, multiple tactics
- From change tactics to change strategies
- From a focus on quality assurance to quality enhancement
- From a focus on fine-tuning of current practice to transforming practice in new directions
- From unscholarly to scholarly
- From context neutral/blind to context- and discipline-sensitive.

This is a very precise description of how I see the trends from the 1990s to the current time. My experiences are formed in a Danish, Nordic, and European context where we started more or less from zero. We had to start somewhere to understand what these new functions would include and the directions they should take. The roles of educational developers have been diverse, and they often act as change agents at universities. This requires respect and acknowledgement from the involved academic staff who undertake the change. Debowski sees this as a new role for educational developers who create a collaborative partnership with academic staff involving the negotiation of interventions (Debowski, 2014).

When I started my career as an associate professor, I was rather quickly involved in the founding of the Centre for Teaching and Learning at Aalborg University in 1994. At that time, there were few educational development units or networks on educational development. Aside from the ICED and the UK organization Staff and Educational Development Association (SEDA), the European Society for Engineering Education (SEFI) was an important place for meeting, exchanging and empowering ourselves.

For a long time in most Scandinavian countries, academic staff development has been compulsory by national law and this has slowly spread to many European countries (Kolmos, 2010). Regardless of trends or how educational development units have supported the transition to a student-centered curriculum or developed strategies rather than tactics, educational development units are management and leadership interventions.

Educational development units must interact at all levels of the university organization, such as attending meetings with department heads, study boards, and deans (Fig. 10.1). The division of expertise varies by university, but when Ruth Graham calls for educational leadership and education of educational leaders, this is frequently left to the educational development units (Graham, 2015).

Funding and, in most cases, strategies must be negotiated with university management, and this means engaging with the management level and carrying out these decisions by interacting with academic staff. With shifting leaders, this requires them to understand and speak the language of leaders.

Trust needs to be created among the academic staff to successfully carry out possible changes. If a relationship between an educational developer and the academic staff has not yet been established, then this will be the first point on the agenda;

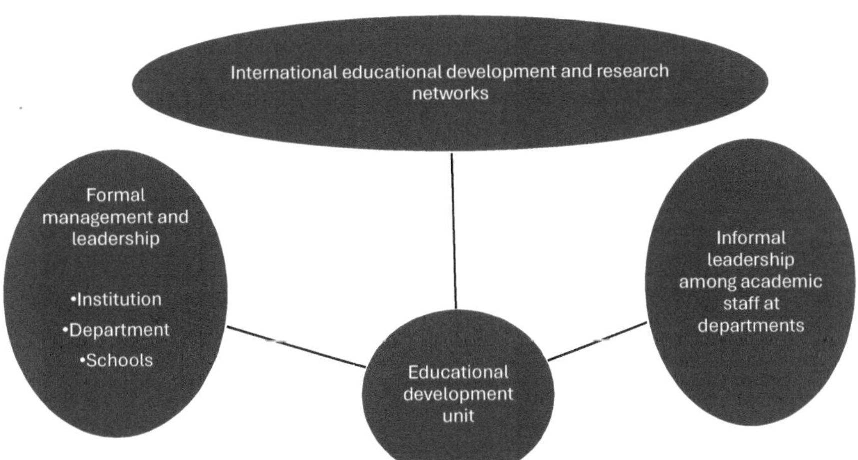

Fig. 10.1 Educational units between organizational relations

otherwise, educational developers will be regarded as outsiders. For that purpose, it is also necessary to speak the language and understand the learning of the content in the discipline and culture. If you know or speak the language in engineering, it is much easier to connect to engineering educators and create relationships and trust.

This is one of the reasons that I, as founding director of the UNESCO PBL Centre, have always tried to identify or educate double-qualified employees, either with a background in engineering/science combined with educational research or in the humanities/social sciences with a learning of engineering thinking and practice. This is a demanding process with many dilemmas both for the organization and the individual researcher.

10.3 Location of the Educational Development Units in the Organization

So where should an educational development unit be in the organization? This is a question that has yet to be answered, and during my career I have seen educational development units start out as decentralized units with research in a department, get moved to the central administration without research, and after some years to become again decentralized. Educational developers are somewhere in between, and there are many creative solutions to the double-binding both to the central and decentralized levels in organizations. Most often there will be a change after a period of time.

There are five core functions for an educational development unit: pedagogical courses and education, educational research and development, evaluation tasks, consultancy for academic staff, and consultancy for management (Table 10.1). These functions are related to both the central and decentralized levels. From an ideal angle, only consultancy tasks have only one organizational affiliation, and even that can be debated.

Many universities have a central unit, and educational developers are given office space within the departments. This highly fluctuating work environment can be difficult to maintain for extended periods of time, such as organizing social and professional relationships at both the central and decentralized levels. On the other hand, this model seeks to combine the relationship between educational developers and

Table 10.1 Functions for educational development units and ideal organizational affiliation

Task	Central	Decentralized
Pedagogical courses, local networks and educations	x	x
Educational research and developments	x	x
Evaluation and research	x	x
Consultancy for academic staff		x
Consultancy for management	x	

departmental academic staff. There is no perfect solution as functions and relationships in the organization will always be fraught with difficulties.

The centralization/decentralization decision also relate to research. Centralized units seldom have research responsibilities, as research positions are in the department and not the administration. This was at the back of my mind during the latest merge process, where the UNESCO Centre for PBL moved from the faculty level to the central level to merge with educational researchers from the other two faculties in an Institute for Advanced Study in PBL. The idea was to create a cross-faculty unit to support PBL in all disciplines—not only in engineering and science. A well-meant thought. In this case, the UNESCO Centre for PBL kept the research positions at the department level, with the ability to also function at the central level. This is also a strange and complicated construction, with it being in the middle but also at the same time related to decentralized levels.

Administration and research cultures do not always align, and as researchers we did not understand the administration culture. An organizational move like this has been seen at other universities and often followed by a reorganization where the units again were decentralized. At some universities, there are both a central unit coupled with decentralized positions. This can work but can be costly.

10.4 Research-Based Development

Whenever there is an organizational restructuring process, research will be at stake as one most often must find the money externally or meet the criteria for internally funded research. With the way many university criteria function today, the units must act as research units and maintain external projects otherwise their importance and recognition in research will decline. The university system will expect development and research close to practice and, if not, it will be hard to understand why this research is necessary. This is a difficult balance because you also must speak the theoretical language of research while also being able to apply your research in practice. Therefore, the research dimension also creates dilemmas for educational developers.

I maintain, however, that educational development cannot function without research, and that educational developers will be drained of knowledge if there is no research base to draw upon personally as well as organizationally. In a research community such as a university, logic and arguments must be grounded in data and research, even for academic staff from various disciplines and, most certainly, for management to make decisions about how to develop a learning culture within a discipline. For academic staff, educational research might give an overview of the advantages and disadvantages of learning methodologies. For management, educational research makes a difference for the decision-making process, as it is always more convincing to base decisions on some kind of evidence.

That engineering education research is an interdisciplinary field and has struggled with recognition and acknowledgement from engineering has been addressed in several studies. In the US version of EER, Beddoes identifies the two main trends of rigor and methodological diversity (Beddoes, 2014). Rigor can lead back to the fact that most of the EER researchers in the US had backgrounds in engineering and were trained in logic, quantitative methods, and a discourse of cause-effect and evidence.

During the 2000s, qualitative methods struggled to be accepted as research and, as the first international associate editor for ASEE Journal of Engineering Education, this gave me some grey hair. Methodological diversity was slowly given more space, and educational researchers became associate editors of engineering education journals. Later on, several articles provided an overview of a spectrum of methodologies—both quantitative and qualitative (Borrego & Bernhard, 2011; Case & Light, 2011).

Even if the field stressed the importance of qualitative methods, it might still be the quantitative methods creating the arguments for change that are acknowledged by the deans and vice chancellors. This also illustrates another research-practice dilemma, where on the research side there is a dialogue with interdisciplinary educational fields, and on the other side the relation to practice with a certain logic and understanding (Klassen & Case, 2022).

So even if you might debate the necessity of research in development, the research methodologies might not only be decided upon by the object/problem of the study but indeed by the considerations of the field of practice, and in this case also the teaching and learning in engineering education. This is something that the EER society seldom speaks about and can be understood as a kind of instrumentalism (De Graaff & Kolmos, 2014).

In more positive terms, usefulness is a concept that has dominated the European Journal of Engineering Education which has undergone a sincere development from being focused on development to engineering education research including usefulness. This process has taken years and some of the ideas for usefulness are to be found in the international networks on development. Historical studies show that the dilemma between theory and practice has always been on the agenda for engineering education (Edström, 2018).

10.5 Working with Resistance

Not only are educational developers caught between contradicting interests and have to learn to navigate in a sea of chaos, working with academic staff can involve both positive and negative reactions, especially if educational developers promote change. I have been involved in the change processes to implement more problem-oriented projects in engineering education, which has not been any easy task and with divergent results.

In a study of an institution's merger with Aalborg University and the resulting changes to its educational model, the resistance was uncovered (Kolmos, 2002). There are always unique circumstances in transformation processes. Either the leaders involved supported the earliest phases of the shift and created a drive, or the change occurred because of deliberate resource cuts, as discussed in Chap. 9, and in this instance the curriculum reform coincided with the unification of two institutions. In such cases, there will always be a big brother institution and a smaller institution. It was a very unequal relationship, with the added danger that it would be closed down if the conditions were not met.

Because the curriculum had to follow the Aalborg PBL model, the staff group that was critical to the reform, chose to look for work elsewhere. Later, we discovered that these academic staff, based on their beliefs and practice, were truly divided into three groups one year after the changes of the educational model (Algreen-Ussing et al., 2000; Kolmos, 2002):

1. A group with a laissez-faire strategy, where the extent, tempo and course content remained unchanged. When the same amount of material was not covered due to the reduced course hours, they simply stopped when they ran out of contact hours.
2. A compression strategy, where the textbook and topics were the same but condensed into fewer contact hours to meet the reduced course time. As a result, the students' workload became significantly heavier.
3. A selection strategy based on reformulating the learning outcomes, such as the central theoretical ideas, and utilizing the relationship to the projects.

The first two strategies are not acceptable, you may say, but the staff might not have the knowledge to act in different ways. The selection strategy could be the most difficult to work with and there are many unknown factors, especially the understanding of the kind of learning outcomes that students can achieve in projects. If you have not had this experience, it might be difficult to plan and formulate learning outcomes. The interviewees stated that they would not 'trust' the new models until they had an overview of the combined educational progression. This would mean that they wanted to see their students get employment and hear employer reactions.

Experiments in the classroom are therefore important pre-phases for any systemic change, together with discussions on new models and difficulties, learning about research in the field, and creating visions in learning circles consisting of peers involved in the transformation. Systemic change is not an individual endeavor but rather a collective cultural change where the pieces will have to fit into new patterns. What is important to remember is that educational developers are just a means for a system decision, so this is not about personal conflicts.

10.5.1 *Loss of Control and Authority*

Educational developers need to be aware that the issue of loss of control and authority is a huge part of the resistance, and a significantly difficult part in a change process. In a student-centered curriculum such as a problem- and project-based curriculum, the relationship between the academic staff and students becomes interesting. Students are, in principle, the owners of the learning process and they direct their own learning within that context. The types of problem the students are expected to work with are normally given in the project proposals. For teachers it is easier to master simple and maybe complicated problems, and more difficult to handle complex and chaotic problems (Fig. 10.2).

For your own self-identity, it is not easy to move from content presentation to facilitating learning processes. When you are used to a content presentation mode, your satisfaction will be dependent on both the delivery and response to the delivery. However, to facilitate collaborative learning where it is difficult to see through students' learning internal processes, it might feel strange and awkward. As you move from simple to complex problems, the problems are no longer defined with clear boundaries. These open and challenging landscapes must be navigated together with the students. As we are on the edge of known knowledge, this also requires more engagement and energy compared to facilitating processes in simple problem-solving ways. Most often this also counts for interdisciplinary projects.

The relationship between academic staff and students will be an ever-balancing dance, both in the planning and design of the learning outcomes and framework for student projects and in the process of facilitation. If the learning outcomes are too narrowly formulated, there will not be many chances for students to make their own decisions. If, on the other hand, they are broadly formulated, then the project facilitators might have less content knowledge in relation to the project.

	Problem	Methodologies	Result
Simple problems	+	+	+
Complicated problems	+	+	?
Complex problems	?	+	?
Chaotic problems	?	?	?

Fig. 10.2 Academic staff's expertise in facilitating types of problems in student projects

The transition to methodical and methodological approaches in PBL is not limited to students' learning of transformation competencies, as described earlier in this book. It is also important for academic staff to assist students with these methods and methodologies and their application and issues throughout the transformation processes. With chaotic problems, both students and academic staff will have to learn fast reaction procedures.

The role of academic staff will therefore be very different. As project supervisor, you should be able to read the students' collaboration patterns and needs in the different phases of the project work and choose the best possible strategies to facilitate a progression towards learning outcomes. You must be able to switch between different strategies, roles and functions depending on the need. Part of the challenge can be met by planning and reflecting on the collaboration with the group, but also lies in being attentive and asking about the students' needs in the supervision situation itself (Kolmos & Holgaard, 2007). This is the art of being able to practice situated project supervision.

In the Danish research literature on project supervision, there is a distinction between product and process as one dimension, and open and control as the other (Olsen & Pedersen, 2008). Students may need answers and guidance, and sometimes they just have a question that helps them to continue on in their process. As academic staff, we must master all types of roles. With this, the educational developers need to have an understanding of academic staff in a change process, or in future processes on how to handle chaos in education. The change from lecturing to supervision and facilitation might provide insights into how to prepare academic staff to frame the handling of chaos.

10.6 Networks as Change Agents

I have described the dilemmas for educational developers and some of the difficulties in their work. There have been two strategies for coping with all the dilemmas: internal networks and strong relations with colleagues, and external international networks with colleagues with similar functions, see Fig. 10.3.

As engineering education research and development is a relatively young field, the first 30 years have been dominated by the establishment of both national and international networks, journals, conferences, common PhD courses, and many other activities. Books could be written about the networks and experiences in the international and intercultural collaboration formed during that time. That is not the intention with this section but it aims to illustrate the kind of research and development network helpful to the institutional educational development units.

In my case, the institutional unit was the UNESCO Chair and UNESCO PBL Centre. The UNESCO PBL Centre was established in 2014, prolonging the original UNESCO Chair in PBL in Engineering Education from 2007 to 2014. There is a history that comes before the decision to apply for a UNESCO Chair and that needs to be told (Enemark et al., 2006; Fink et al., 2003; Kolmos et al., 2007). We had

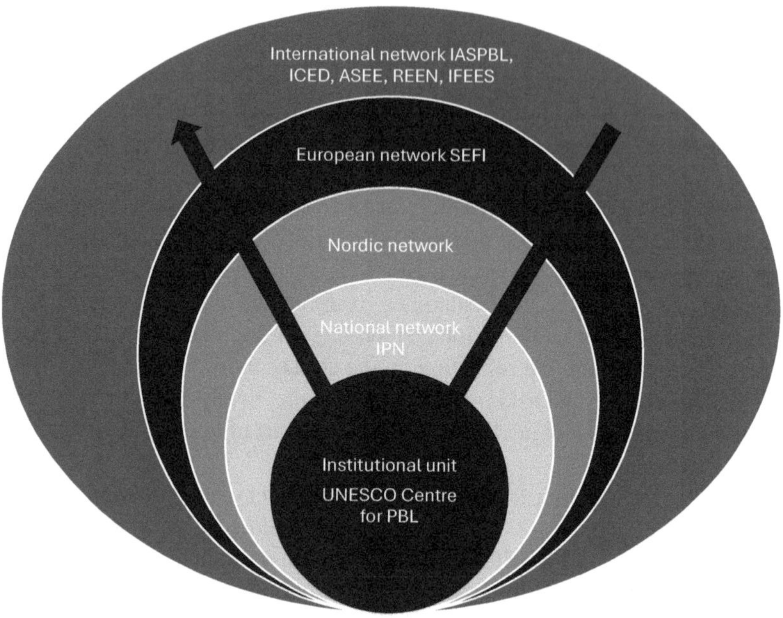

Fig. 10.3 National and international networks

many visitors during the 1990s, especially from the Netherlands and Scandinavia, and there was an increased interest from both the UK and the US. There was, however, a rising awareness that Aalborg University had something that attracted international interest. Compared to other universities, research profiles took time to become established; however, the pedagogical model was a distinctive characteristic for Aalborg University and to disseminate it, research and publishing helped to get the model known.

The functions of the Chair and the Centre were similar, having the following objectives:

> 1) to create a global society of practitioners, researchers, experts and institutions within the field of Problem-Based and Project-*Based Learning (PBL) and Sustainability in Engineering and Science Education.*
>
> 2) to establish international research and doctoral training on PBL and sustainability in Engineering and Science Education.
>
> 3) to provide global formal education and training for academic staff and students and to disseminate and exchange knowledge and support among society members in terms of how to change engineering and science education for regional sustainable development.
>
> 4) to give HE institutions and governments open access to a body of knowledge, education, training, and other resources in order to facilitate PBL and Sustainability in Engineering and Science Education.

Besides these objectives, we had the internal responsibility of introducing the engineering students to PBL and organizing internal educational activities for academic staff. Budget-wise, it was the internal activities that provided the significant income to the Chair/Centre; however, it was the outreach combined with the institutional international activities that contributed to the acknowledgement internally.

Not all institutional units can do the same, and it has been unique for us to have the experience and research on PBL, which became very popular and was in demand by the international engineering education communities. We started out with one professor and a smaller group of colleagues who were primarily responsible for the PBL teaching in the first year and growing to 35 academic staff members in 2023 when I handed over the leadership to my successors.

10.6.1 PhD Students and Training the Trainers

One of the core strategies for educational change globally was to educate PhD students. It would not be possible to run training courses and facilitate change at all the universities that requested our help. Furthermore, educational development in China, India, or South America would be quite different due to the cultural diversity. We have educated PhD students from China, India, Pakistan, Malaysia, Thailand, Colombia, Spain, Italy, Sweden, and the USA. A huge part of the job as Director for the UNESCO Centre for PBL has been to find available funding and collaborate with the institutional cultures, which partly financed the PhDs. The strategy was to let the PhD students collect data at their home institutions to be combined with theories on student-centered learning and active learning methodologies.

As Aalborg University is a full-scale PBL institution (see Chaps. 3 and 4), this pedagogical system could never be adapted to colleges in e.g. India, China, Thailand or any other place since adaptations and considerations of national and local legislation would be necessary. I am proud to say that in 2026 we find variations of PBL in nearly all the mentioned countries, with the PhD students playing a role in this development. I have earlier in this book asked if it has been worth the effort we have made? On a personal level, this has involved a great deal of travelling with both positive and negative aspects to my life, and as an academic I would say that the progress is worth the personal and energy costs.

10.6.2 International Research Symposium on PBL (IRSPBL)

One of our objectives was to create a global network and we started with a series of conferences, which we called the International Research Symposium on PBL (IRSPBL). The first IRSPBL was held in 2007 at Aalborg University, followed by biannual symposia in Australia, the UK, Malaysia, Spain, Colombia, China, India, again at Aalborg (virtual because of Covid-19), the US, and finally Africa, which I had in the pipeline when I handed over leadership (Fig. 10.4).

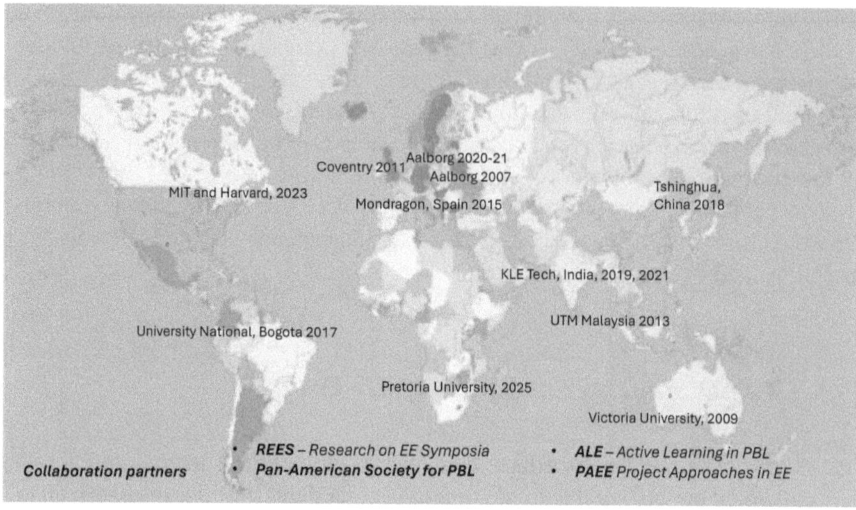

Fig. 10.4 IRSPBL 2007–2025

This was a small community, and the first symposia attracted around 120–150 participants, most of whom came from the local area. However, the fact that we moved around and teamed up with local universities provided a decentralized anchor, with huge thanks to the involved institutions. These chosen institutions were the ones going through a transformation process. In the beginning the research approach was not mature, being more case descriptions with evaluations than research. The research level has increased during the years and this is now similar to the research level in EER (Arana-Arexolaleiba et al., 2017; Davies et al., 2011; Du et al., 2009; Kolmos et al., 2010; Mohd-Yusof et al., 2016). One takeaway at the conferences was that experiments should be documented, and further development should be based on research.

Reflecting on the years of energy invested in this movement, I dare to conclude that we made a difference in several aspects:

- Firstly, there has been increased research and increased levels of scholarship. Being co-host for nearly 10 IRSPBL symposia, I have seen how it has moved from case descriptions to referenced literature and regular studies on students' learning, and it has been a pleasure to follow this development of moving through levels of scholarship (Boyer, 1990; Trigwell et al., 2000). A whole research society has been built up around PBL.
- Secondly, we have also seen that research-based development at some institutions has had an impact on other institutions' decisions to change their education.
- Thirdly, we have created a network that can be an inspiration for change, with institutions involved at all levels considering a change process to experience and further develop it. We approached other international networks that have strengthened the trend towards moving to more student-centered learning. As the IRSPBL

> was a rather small network, we have always tried to reach out to other smaller networks like the Active Learning in Engineering (ALE), Project Approaches in Engineering Education (PAEE), the international networks Pan-American Society for PBL (PAN-PBL), and Research in Engineering Education Network (REEN). The conference activities were organized as back-to-back activities, which always attracted more participants.

The network has consisted of rectors, deans, academic staff, educational developers, and researchers. As such, we have never purposely reached out to the top management at the institutions, but we have managed to attract them because we are known for having something to offer. Not only was the UNESCO PBL Centre located at one of the pioneering institutions, which indicated a great deal of practice experiences, but the research component has continuously developed from the 2000s until the present time.

10.7 National, Nordic, European and International Networks

We have been involved in other networks since 2007 with the first UNESCO Chair in PBL in EE. The Danish engineering institutions created a national partnership within the field of academic staff development. This National Pedagogical Network for Engineering Education (IPN) was founded in 1996 with the aim of ensuring the quality of the pedagogical and curriculum development of all Danish engineering education (Vinther & Kolmos, 2002).

The IPN initiated Nordic collaboration and later a Nordic network on EER was established (Edström et al., 2016; Edström et al., 2018; Vinther & Kolmos, 2002). There is the Nordic Council that provides diverse funding programs, one of the arguments for establishing the Nordic network; however, another reason was the common language, culture, geographical locations, and closeness. Several PhD courses have been held at the Nordic level.

Several of the same arguments, except for language, could be valid for the European Society for Engineering Education. Around 2008, I became the founding chair for the working group on engineering education research. I stepped back in 2009 to take over the presidency of SEFI, which is the organization for engineering education in Europe and a core actor for development of engineering education. SEFI has always worked with the sister organization American Society for Engineering Education (ASEE).

As described in Chap. 6, the Conceive, Design, Innovate and Operate (CDIO) group has been active at the international scene and shares many ideas with both CBL and PBL (Crawley et al., 2007; Crawley et al., 2014; Edström & Kolmos, 2014). CDIO is an international network closely connected to the development of educational practice and is one of the organizations that plays a significant role in the move to more student-centered and active learning. The last network I will mention here is the International Federations of Engineering Education Societies

(IFEES), which, as the name indicates, consists of a network among the regional and international engineering education societies and has a strong voice to deans and vice-chancellors.

All of these organizations and networks have been involved in facilitating change in engineering education, and indeed there are many more. I have not mentioned much about accreditation in this book, and there is no doubt that the Washington Accord and the US ABET criteria have had important roles in the regulation of higher education especially in countries with a greater development potential from lecture-based curriculum. However, a network cannot stand alone, and it is the combination of networks that can disseminate ideas and, in the end, influence practice.

10.8 Reflection 10 on Learning to Live-with Chaos

This chapter is based on experiences that create a change driver for educational development and educational networks. Maybe it is a detour, nevertheless as a change agent, you will encounter significant resistance and require support and peers to share and evaluate your experiences with.

The transition from a teacher-controlled to a more student-centered system has been and continues to be a significant change. It affects how we learn and shape ourselves to become active and collaborative learners who take responsibility for our own decisions. The PBL system prepares candidates to give and take, receive and create, and reflect and act.

Perhaps incorporating chaos into different educational variations will not result in such dramatic changes. It will include a content dimension, as well as influencing risk management learning and increasing interdisciplinarity. Specifically, risk management will necessitate an interdisciplinary approach that considers both technique and its impact on the people who use it. It will also include a new approach to learning generic competencies as methods, as well as the integration of emotional maturity development. If an educational system is already on the path to more student-centered learning, incorporating it into the existing curriculum should be relatively simple.

This could prove to be a particularly difficult transition since it involves establishing a new academic culture, and the first step is to discuss it—to discuss our anxieties and hopes for the future, as well as how the climate and political situation affect us as individuals and communities, and to discuss technological solutions, risks, and how to respond in critical situations. It may be difficult to create a new discourse which is not only concerned with how you speak about things but where you create new culture with habits and actions, a culture in which risky situations are clearly taken into account.

There are educational development units, networks, and opportunities to influence current practices. It is a matter of raising the issues and thinking them through. Broaching concerns about students' learning and competencies, as well as directing them toward a PBL system, was a difficult task. There is no doubt that establishing the UNESCO PBL Centre and receiving recognition from international organizations such as UNESCO has been a catalyst for change in engineering education around the world. This center has served its purpose of raising issues, documenting effects, pointing in the direction of PBL implementation strategies, and creating networks.

There may also be a need to address the issue of how to live-with chaos. And for this type of change, a network is one way to move ahead together with research. We need to exchange ideas of how to deal with learning to live-with chaos. Is it in separate courses? Is it a cultural change and integration in the existing curriculum? Is there any scientific content we need to focus on? This leads to my last chapter on living-with chaos.

References

Algreen-Ussing, H., Keiding, T. B., & Kolmos, A. (2000). *Pædagogisk omstilling, læringsopfattelser og organisatoriske rammer* (2. oplag. ed.). Videncenter for Læreprocesser Pædagogisk Udviklingscenter.

Arana-Arexolaleiba, N., Guerra, A., Kolmos, A., De Graaff, E., & Lima, R. (2017). The learner in engineering education [editorial]. *International Journal of Engineering Education, 33*(3), 940–941.

Beddoes, K. (2014). Methodology discourses as boundary work in the construction of engineering education. *Social Studies of Science, 44*(2), 293–312.

Borrego, M., & Bernhard, J. (2011). The emergence of engineering education research as an internationally connected field of inquiry. *Journal of Engineering Education, 100*(1), 14–47.

Borrego, M., Streveler, R. A., Miller, R. L., & Smith, K. A. (2008). A new paradigm for a new field: Communicating representations of engineering education research. *Journal of Engineering Education, 97*(2), 147–162.

Boyer, E. L. (1990). *Scholarship reconsidered: Priorities of the professoriate*. ERIC.

Brunhaver, S., Yasuhara, K., Case, J., Newstetter, W., Turns, J., Carberry, A., Allendoerfer, C., Finelli, C., Sheppard, S., & London, J. (2018). *Meet the Engineering Education Pioneers—Panel & Roundtable*. 2018 IEEE Frontiers in Education Conference (FIE),

Buckley, J., Wallin, P., Matemba, E., Power, J., Mohanty, A., & Bombaerts, G. (2023). The future of engineering education research. *International Handbook of Engineering Education Research, 1*, 711–729.

Case, J. M., & Light, G. (2011). Emerging research methodologies in engineering education research. *Journal of Engineering Education, 100*(1), 186–210.

Crawley, E., Malmqvist, J., Ostlund, S., Brodeur, D., & Edstrom, K. (2007). Rethinking engineering education. In *The CDIO approach* (Vol. 302, pp. 60–62).

Crawley, E. F., Malmqvist, J., Östlund, S., Brodeur, D. R., & Edström, K. (2014). The CDIO approach. In *Rethinking engineering education* (pp. 11–45). Springer.

Davies, J., de Graaff, E., Kolmos, A., Davies, J., de Graaff, E., & Kolmos, A. (2011). Introduction. In A. Kolmos (Ed.), *PBL across the disciplines* (pp. 1–2). Aalborg Universitetsforlag.

De Graaff, E., & Kolmos, A. (2014). Innovation and research on engineering education. In *Handbook of research on educational communications and technology: Fourth Edition* (pp. 565–571). Springer New York. https://doi.org/10.1007/978-1-4614-3185-5_44

Debowski, S. (2014). From agents of change to partners in arms: the emerging academic developer role. *International Journal for Academic Development*, 19, 50–56. https://doi.org/10.1080/1360144X.2013.862621

Du, X., de Graaff, E., Kolmos, A., Du, X., de Graaff, E., & Kolmos, A. (2009). PBL–diversity in research questions and methodologies. In A. Kolmos & A. Kolmos (Eds.), *Research on PBL practice in engineering education* (pp. 1–7). Sense Publishers.
Edström, K. (2018). Academic and professional values in engineering education: Engaging with history to explore a persistent tension. *Engineering Studies, 10*(1), 38–65.
Edström, K., & Kolmos, A. (2014). PBL and CDIO: Complementary models for engineering education development. *European Journal of Engineering Education, 39*(5), 539–555. https://doi.org/10.1080/03043797.2014.895703
Edström, K., Kolmos, A., Malmi, L., Bernhard, J., & Andersson, P. (2016). A bottom-up strategy for establishment of EER in three Nordic countries—The role of networks. *European Journal of Engineering Education*, 1–16. https://doi.org/10.1080/03043797.2016.1190956
Edström, K., Kolmos, A., Malmi, L., Bernhard, J., & Andersson, P. (2018). A bottom-up strategy for establishment of EER in three Nordic countries–the role of networks [article]. *European Journal of Engineering Education, 43*(2), 219–234. https://doi.org/10.1080/03043797.2016.1190956
Enemark, S., Kolmos, A., & Moesby, E. (2006). *Global network on engineering education research and expertise in PBL: The profile of UCPBL.*
Finelli, C. J., Borrego, M., & Rasoulifar, G. (2016). Development of a taxonomy of keywords for engineering education research. *European Journal of Engineering Education, 41*(3), 231–251.
Fink, F. K., Enemark, S., Moesby, E., & Kolmos, A. (2003). *Global network and global Centre for problem based learning.*
Gibbs, G. (2013). Reflections on the changing nature of educational development. *International Journal for Academic Development, 18*, 4–14. https://doi.org/10.1080/1360144X.2013.751691
Graham, R. (2015). *Does teaching advance your academic career?: Perspectives of promotion procedures in UK higher education.* Royal Academy of Engineering.
Jesiek, B. K., Newswander, L. K., & Borrego, M. (2009). Engineering education research: Discipline, community, or field? *Journal of Engineering Education, 98*(1), 39–52.
Klassen, M., & Case, J. M. (2022). Productive tensions? Analyzing the arguments made about the field of engineering education research. *Journal of Engineering Education, 111*(1), 214–231.
Kolmos, A. (2002). Facilitating change to a problem-based model. *International Journal for Academic Development, 7*(1), 63–74. https://doi.org/10.1080/13601440210156484
Kolmos, A. (2010). Danish faculty development strategies. In M. Frenay (Ed.), *Building teaching capacities in higher education.* Stylus Publishing.
Kolmos, A., De Graaff, E., & Du, X. (2010). Guest editorial: Methodology for the study of PBL in engineering education [editorial]. *International Journal of Engineering Education, 26*(3), 508–510.
Kolmos, A., Enemark, S., Du, X., Moesby, E., Kolmos, A., Enemark, S., Du, X., & Moesby, E. (2007). Czy warto wprowadzic problem i project based learning? Innowacje w edukacji akademickiej, *1*, 99–112.
Kolmos, A., & Holgaard, J. E. (2007). Situationsbaseret projektvejledning. *Dansk Universitetspædagogisk Tidsskrift, 2*(3), 54–62.
London, J., Yasuhara, K., Carberry, A., Allendoerfer, C., Ayela-Uwangue, A., Cruz, S. N., Lee, E., Huerta, M., Abhyankar, R., & Huang, W. (2021). The Pioneers' stories as a tool for introducing graduate students to the engineering education research community. *Advances in Engineering Education.*
Mohd-Yusof, K., De Graaff, E., Kolmos, A., & Phang, F. A. (2016). Selected papers from the international research symposium on problem based learning, IRSBL 2013. *International Journal of Engineering Education, 32*(1), 305–307.
Olsen, P. B., & Pedersen, K. (2008). *Problem-oriented project work.* Samfundslitteratur.
Trigwell, K., Martin, E., Benjamin, J., & Prosser, M. (2000). Scholarship of teaching: A model. *Higher Education Research & Development, 19*(2), 155–168.
Vinther, O., & Kolmos, A. (2002). National Strategies for staff and faculty development in engineering education in Denmark. *Global Journal of Engineering Education, 6.*

Chapter 11
Learning to Live-With Chaos

11.1 Introduction

This is the final chapter of the book. I have enjoyed writing it and putting PBL into a new perspective and have raised many questions along the way. I have outlined some possible but maybe unclear shapes for my prospects. Writing for me is a cognitive and deep emotional process, where it is in writing and reading that ideas and more profound understandings take form.

We need many more ideas of how to respond to chaos, and this is the real purpose of putting together all these PBL perspectives in the shadow of chaos. A first step is to acknowledge that chaos exists within society, the current political landscape, technological and climate-caused situations, and within the lives of individuals. There are many levels at which chaos can exist, and a few years ago I might have said that facilitating the individual lives does not belong in engineering education. But perhaps I was wrong. Perhaps the increasing levels of societal chaos, along with the rise of digitalization, AI, and vast amounts of information, indicate that the future of higher education should not only focus on scientific knowledge and competencies but also emphasize the importance of fostering stronger individual and collaborative resilience.

I will sum up my messages here in this final chapter and add some perspectives. My hypothesis has been that a PBL curriculum can be a good starting point for adding elements that address the chaos agenda. However, as I have reasoned previously, the curriculum will need some adjustments and the integration of new elements.

In Fig. 11.1, I have made an overview of the core elements, which I will address here in this final chapter. For both the complex and chaos modes, there are potentials from a PBL approach related to four curriculum elements:

A. Kolmos, *Engineering Education for Interdisciplinary Futures*, Emergent Directions in Engineering Education 2,
https://doi.org/10.1007/978-3-032-32540-2_11

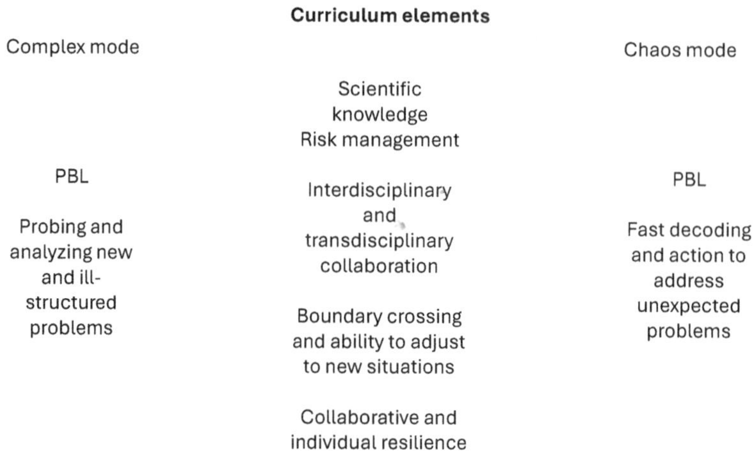

Fig. 11.1 Core elements in a curriculum change to respond to chaos

1. Scientific knowledge and risk management
2. Interdisciplinary and transdisciplinary knowledge and collaboration
3. Boundary crossing and adjustment to new situations
4. Collaborative and individual resilience

Finally, I will reflect on the change strategies for new institutions on the move to more student-centered learning.

University education is fundamentally about acquiring scientific knowledge, but many pedagogical and educational approaches often overlook this. The transdisciplinary collaboration across educational levels could have been developed much more during the book with respect to collaboration between academics and non-academics. At least the arguments are in place.

11.2 Chaos and Complexity Modes

Let me start with chaos. This book has had many titles. I got started by reading Ginie Servant's work *Pedagogies of Collapse*, and I was quickly able to connect 'collapse' to an understanding of chaos (Servant, 2024). Chaos will occur before collapse, and there will be opportunities to maintain hope while establishing constraints to gain some control over chaotic situations. If we consider climate change to be an example of chaos, we, as academics, will deal with conditions beyond our control. Climate change itself is a very complex topic and consists of numerous variables, all of which influence each other. To talk about chaos can be a bridge between collapse and complexity (see more on this in Chap. 2) (Snowden & Boone, 2007).

Dave Snowden's insights into chaos and complexity have inspired me. In Snowden's framework, it may be 'unorder' or 'disorder' that I address, as there is some logic in chaos. Unorder or disorder is characterized by not knowing which domain we are in (Kurtz & Snowden, 2003). For example, the climate models do predict some kind of chaos based on identified variables. Even if we cannot predict the exact climate development, the scientific society has identified the variables. Snowden emphasizes that when we are dealing with a disordered domain, we do not necessarily have the knowledge of the variables behind it.

Regardless of the conceptual understanding, climate change is serious, and it is the largest threat against our life on earth. We have for a long time thought that technology and smart engineers will find solutions to the climate problems. This is not only a matter for engineers and scientists; it is also a matter for politics and regulation. And it is an example of legislation and political decision making that relatively few politicians are interested in, since we will have to slow down economic growth and change existing economic structures. We are rapidly approaching a tipping point, and it is a disaster that our elected politicians, as well as the people who vote for them, are so shortsightedly ignoring. I believe that climate change should be the basis for all political decisions.

We expect climate chaos, but we do not know or understand the full consequences. Today, countries are attempting to incorporate climate solutions into city planning by integrating flooding strategies, and some new home constructions are designed with zero-energy features. We see renewable energy sources growing quickly through the use of wind, sun, and water. The reuse of material is increasing, and new fuels with less CO_2 emission are slowly coming into the market. Sales of electric cars dominate in some Western countries; however, this growth also creates new challenges that need to be addressed, such as the issues related to batteries. All these trends are a preparation for a warmer climate. What if, for example, we are preparing ourselves for the wrong scenario? It could be that the Gulf Stream collapses, and then Scandinavia, for example, will have a much colder climate.

There are different types of chaos ranging from smaller periodic situations to longer-lasting ones. Everyone dealing with technological development and innovation will recognize the shorter periodic chaos that can exist, for example, in work processes, scientific experiments, writing processes, the combination of new devices in a bigger system, and the impact on humans. Chaos is inherent in our technical growth and maintenance, scientific research processes, and daily lives, and it may also be an important source for creative thinking and innovation due to the disruption that a chaotic situation can create. In this book, I argue that we tend to shy away from discussing chaos due to our belief that short-term chaotic situations typically resolve quickly. Therefore, we do not include it in education.

In engineering we have started to introduce complexity into education and prepare students to probe, analyze, and respond to complex problems through methods such as interdisciplinary learning. Complex problems have been highlighted in the recent versions of the ABET criteria (ABET, 2026); however, they seldom become part of engineering education, as this is difficult. As complexity covers a large number of variables and will include interdisciplinary approaches, which are complicated to manage within the educational frameworks.

Deep knowledge and understanding of one's discipline are required, and interdisciplinary collaboration will also involve gaining a broader perspective by understanding other fields of knowledge. You can argue from a sequential point of view that students need to understand the deep knowledge in a discipline. However, they also need to know how to collaborate across disciplines.

Even though complexity and chaos are intertwined, as you will experience periodic chaos in complexity, these are very different reaction modes. Complexity can take place with reading, thinking, experimenting, testing, collaborating, discussing, and writing, all of which are within the realm of probing. Reflection and learning precede action, allowing for well-thought-out, well-argued, and well-planned actions. In chaotic situations, you will not have the time for this; it will require an immediate response to get an overview of any unexpected situation (Kiehl, 2023).

What is important is that immediate actions rely on existing knowledge. Your response will depend on your knowledge and methods. In reflecting the learning theories, Kolb's learning circle has an action mode included via the experimentation, which is not the same as a reaction to chaos but facilitates the process of transferring theoretical knowledge to practice.

One of the messages in this book is not a question of learning less; it is a question of adding a dimension of using the theoretical knowledge in real situations—not only in an analytical perspective but in a practical situation (Pogue, 2021). In order to react quickly, you will have to read the problem situation quickly, and this is where PBL can be an advantage, although this is not how it is practiced, and I will return to this later.

11.3 Living-With

Then there is the 'living-with' in the titles of this chapter and the reflections. Living-with is an inspiration from Verlie (see Chap. 2) (Verlie, 2019). In one of the first title proposals, it was 'Coping,' which had a more dramatic sound than 'Survival'. Perhaps we are facing survival this time around. We are not motivated by just being negative, we are already dealing with chaos and fear at different levels.

We have had nuclear disasters like Chernobyl and Fukushima. In Denmark, the Barsebäck Nuclear Power Plant just across the water from Copenhagen used to cause us constant concern with a possible collapse. During the 1970s we were against nuclear power, and there were huge demonstrations and a strong movement. Denmark does not have nuclear power today. We already have learned to live-with some possible chaotic situations.

The younger generation also need to learn to 'live-with' the chaos and anxiety of the future at a societal, collaborative, and individual level, and one way of taking the first steps is to start talking about how we live-with chaos without becoming passive citizens but with the ability to act. And we must be aware that the decisions we make today will form the future.

Regardless of whether we are living-with chaos or collapse, we are dealing with a generation of young people who have anxiety for the future. Perhaps our knowledge of this issue stems from our research efforts. We did not survey young people concerning these matters 40 years ago. In most of the present studies, the anxiety is mostly related to climate and war. This will influence the personal lives and values that the next generation fosters.

We will all experience chaotic situations in our personal lives. Nearly every generation has experienced chaos at the political or economic level, the numerous wars throughout history, refugee crises, and lack of food. Living-with covers the fact that chaos will always exist for periods of time, and we will not get through life without experiencing it. During my childhood, we had the Cold War and the threat of the Russians, and this is the same today. When we cleared up the house after my father passed away, we found machine guns and ammunition in the attic, as he wanted to be certain he could defend his family. He had experienced both the First and Second World Wars, and it was integrated into his bones that we should be able to defend families and friends.

Other personal challenges, such as divorce or the loss of dear friends and family members, can also create chaos in people's lives. It is not the task of any educational system to prepare students for such situations; however, the educational system needs to start talking about chaos at different levels and how to handle it. Students should be empowered to handle both complexity and chaos, using scientific thinking as their foundation.

11.4 The Potential of PBL

As described in Chaps. 3, 4, 5, and 6, there are many favorable intentions and outcomes of PBL and the learning principles that underpin it, e.g., learning to analyze and solve problems, relate theory and practice, and to collaborate. These are all elements that are useful in learning to handle chaos.

The type of problem that students work on varies with different PBL curricula. As already mentioned in earlier chapters, the problem can be one that is narrow and disciplinary, open and authentic, theoretical or practical, or it can be a wish or an interest. The problem initiates the students' learning process, and throughout the entire project, they continually refer to the problem and its statement. Has it been formulated correctly? Is this the problem that binds together the methodology, theory, and findings? Could there be other angles or perspectives? Have we forgotten anything? We often hear these reflective questions during the facilitation sessions of students' projects.

The students learn to identify problems, issues, and challenges, and they have time to do so; for example, when they go to a company and analyze the efficiency in a production line with the hypothesis that it is the packing that causes delays, they might realize the problem was actually later down the line of logistics. These students have time to analyze the problem and identify the right solution. Or when students from city planning, land surveying, electronics, and civil engineering work on the implementation of a new electronic traffic regulation system based on adaptive technologies, which can be quite a complex challenge, they have time to analyze the traffic situation, develop prototypes, and test before coming up with recommendations.

Time is of the essence in a chaotic situation, as a swift response is necessary to establish constraints and limit any damage. So how can a PBL curriculum train students in rapid problem identification? My recommendation is that students will obtain further training in problem identification, partly by starting to compare the various problem types they work with, and partly by reflecting the initial understanding of a problem with the final identified problem.

Reflection on variation is one of my key theoretical contributions to PBL. What type of problem should the students work on? Even if we are dealing with authentic problems, the students will solve them by analyses, theories, and methodologies. They will not experience a situation requiring immediate reaction.

It seems possible to teach immediate reaction in workshops using different cases, and this leads me to the theory-practice dimension in PBL. Students learn to apply theory and methodologies to real-life problems; however, this is still at a theoretical level because they do not learn to implement or react to the problems. The CDIO framework, as described in Chap. 6, stands for Conceive, Design, Innovate and Operate. However, operate is mostly about building prototypes, which is beneficial for learning engineering but not sufficient to understand operating a real engineered system.

Reacting at fast speed are missing in PBL. I don't think we should add 'doing components' to every project, as it may not be relevant. However, students need to be aware and capable of fast speed problem identification and response. This could be developed in workshops.

In real chaos situations, there might be a need for both academic knowledge and craftsmanship. Therefore, it is important to integrate the transdisciplinary aspect along with the interdisciplinary aspect into education. Both interdisciplinarity and transdisciplinarity should be elements in a PBL curriculum, which is not always the case, as I have discussed in Chap. 7, and it is necessary to create different types of interdisciplinary projects.

The transdisciplinary part, understood as the crossing between academia and non-academia, is an important element of learning for students. What happens if we have a power cut, which happens occasionally, and people are stuck in an elevator? There is a need to look into the energy supply, and this can be both an academic and a technician's competency; there is definitely a need for technician skills to get people out of the elevator or other places. The learning of languages and ways of thinking and acting across educational levels will therefore be valuable in such situations.

A one- or two-day workshop can make a difference for students to start thinking in the direction of what to do when any short period of chaos occurs. For longer periods of chaos, a greater number of factors must be considered in the learning process. Climate-caused chaos is not something one or two people can respond to and will require an effort from many people collaborating across several boundaries. This might be difficult to get integrated into education as such, although the learning of reactions to shorter periods of chaos can be seen as a preparation for the longer chaotic situations.

PBL has strong assertions in the learning of collaborative skills, and this might be one of the strongest benefits. However, reflection on the learning of collaborative competencies is often left out of the curriculum in engineering education, as this is outside of a technical profession, and not many engineering colleagues know what and how to integrate reflection into the curriculum.

11.5 Next Steps

There are potentials in a PBL curriculum in relation to learning to live-with chaos; however, as I have argued in the previous section, PBL does have some pitfalls: the variation of analyzing problems, the relationship to practice and actions, and, not least, the interdisciplinary and transdisciplinary learning. There are advantages concerning collaboration.

Organizational and educational stagnation are part of all systems, as written about in Chaps. 9 and 10. With the rapid growth of society, the educational system must respond at a similar pace. Well-established PBL courses and curricula may face at least as many challenges in altering the systems, students' learning outcomes, teaching and learning methodologies, and culture as traditional institutions do when transitioning to more student-centered learning methods.

If I summed up the curriculum elements that I believe are necessary for responding to the new societal agendas of chaos, then I would point to the four elements mentioned in Fig. 11.1: scientific knowledge and risk management, inter- and transdisciplinary knowledge and learning, boundary crossing, and resilience.

11.5.1 Scientific Knowledge and Risk Management

You do not need a PBL course or a PBL curriculum to teach science and engineering in an open-question approach. For example, inquiry-based learning does not indicate the organization of the learning process; rather, it emphasizes the cognitive process as an inquiry process with open questions and learning to ask questions.

Learning to ask questions is not necessarily the same as problem analysis, as the latter involves, for example, stakeholder identification and sometimes power relations among stakeholders. The inquiry-based approach can be the scientific questions without any relation to the involved context. However, to learn to ask questions stimulates curiosity and awareness of the context.

Risk management is about learning to avoid loss or to minimize negative impacts (threats), and risks might be built into standardized systems. Risk management is often not an integrated part of students' projects and engineering work, although there are accreditation criteria specifying that risk should be part of design projects. It is mostly related to economic risk rather than safety, even if safety is the second most important level in Maslow's hierarchy of human needs (Meyer & Reniers, 2022). Perhaps there is still room for development in this area even if risk management has been part of the ABET criteria since late 2010, which is ahead of engineering education practice concerning both risk and complexity.

There are safety courses in relation to laboratory activities that introduce experiments and other activities. These are mostly to protect the students from any disasters in the lab. Risk management includes the question of how to handle uncertainties, and this can be engineered in a structured way. In this book, I argue that risk management will be an important component at a general level and indeed that there is a need to update students' preparedness for possible risk and chaos situations.

Risk management is most often an interdisciplinary activity, since human factors are involved and there is an embedded ethical dimension. Findings from a review study indicate that risk management often is anecdotal or illustrative in many of the articles dealing with coupling ethics and risk management (Guntzburger et al., 2017). The concept of technology assessment covering the analyses and impact of technique, knowledge, organization, and the product on society also points in the direction of including risk management (Müller et al., 1984).

Integrating risk management more systematically in engineering education could provide the students with an overview of different cases and strategies for actions at a knowledge level. Students learn to identify risks, which helps them develop a preliminary understanding of how to find solutions in chaotic situations. This is an important element that could be easy to integrate into PBL as part of the students' projects. The resistance in the systems lies in the interdisciplinary aspects, which are embedded in a risk management approach. Courses on risk management can also be established in a traditional course-based system.

11.5.2 Interdisciplinary and Transdisciplinary Knowledge

At this point in the book, it should not be a surprise that I would argue that inter- and transdisciplinary learning should become compulsory elements in the curriculum. Students should learn how to collaborate across disciplinary and educational boundaries. As indicated in the previous chapter, risk management very often includes an inter- or transdisciplinary aspect.

It should also be clear that I advocate for collaboration across disciplinary boundaries and not necessarily to establish new programs. The main purpose is that interdisciplinary programs serve another purpose, which is to educate boundary workers and, if there is a need from society, to educate candidates for new functions. As indicated in Chap. 7, establishing new interdisciplinary programs and qualification profiles creates both a push and a pull in the labor market, which can lead to new innovations.

Inter- or transdisciplinary learning does not always require PBL, though collaborative features are important learning objectives that are difficult to attain without projects. Other active learning methods can be employed for this purpose, which may involve shorter periods of different classroom activities; in elective curriculum systems, it may be easier to address this because students must earn credits from various disciplines and can select courses where they will be grouped with peers from those disciplines. In this regard, the US system has a significant advantage by combining a specific number of mandated credits with credits from elective courses (Habbal et al., 2024).

Inter- and transdisciplinary learning across existing boundaries, as well as facilitating students' acquisition of interdisciplinary competencies, should not happen at random; there is a progression in learning interdisciplinary competencies through experience in order to collaborate smoothly and with trust.

Transdisciplinary activities can be short-term workshops, longer-term projects, or something in between. What is important are the students' reflections on crossing disciplinary and educational boundaries to identify and solve problems quickly.

11.5.3 Boundary Crossing, Transformation and Generic Competencies as Methods

The number of competencies is growing, and rather than adding more, I see boundary crossing as one of the core generic competencies for the future. Boundary crossing can be defined as the learning of crossing any boundary, such as different disciplines, educational backgrounds, cultural differences, the transition from education to work, moving from one project team to the next, or from one problem to a new problem situation.

These examples are dissimilar, and boundary crossings are not the same in interdisciplinary collaboration as in cultural collaboration. I mentioned in Chap. 8 that cross-cultural collaboration can be much more difficult than interdisciplinary collaboration, where there may be a shared cultural value. It will be difficult to cross both disciplinary and cultural boundaries at the same time.

While the specific contexts will influence boundary crossing, the overall steps of identification, coordination and othering, reflection, and transformation remain. This is one of the reasons that I argue that generic competencies should be learned as methods rather than as specific skills. Learning generic competencies as methods involves the integration of transformation from one context to a new one. Some would call these meta-competencies, the abilities to aid in developing other competencies, which I have described in earlier publications (Habbal et al., 2024). I am no longer sure that we need to have two levels, as the learning and application of competencies always will be a fluctuating process.

By emphasizing the learning of methods, it indicates that competencies will never be stable or remain the same. Situations and learning processes are constantly changing, leading to variations in all variables across different contexts. When learning to do management and work organization, the purpose, tasks, resources, and timeline are all interdependent. Therefore, variation and adjustment to the specific situations are already woven into the competencies from the start.

The transformation aspect should be built in since competencies will be learned in one situation and maybe learned deeper in a slightly different or new situation. However, there will always be differences and variations in the situations. Variables will change, and there will be differences from one day to the next even for a project team working on a common project.

Learning generic competencies as a method involves reflecting on variations in situations and conditions. The awareness of variables in specific situations will help facilitate the transformation processes, for example, the learning of project management in teams of four versus the project management system in a team of seven.

11.5.4 Collaborative and Individual Resilience

Building up resilience in education is not a new discussion. Building resilience has been on the agenda for some years, and in this book, I have addressed the topics of reflection, the learning of generic competencies, emotions, agency, and participatory learning, all of which influence both individual and collaborative learning. These elements and many others will influence the building of both the self and the inner robustness to handle new, unknown, and unexpected situations. In education, especially in the participatory dimension, it is important alongside what others might call self-directed learning or co-design. To build resilience, students need to learn how to make independent decisions on their own and in collaboration with others.

Decision-making is part of agency and creating trust in oneself, not only to do it but also to reflect on the experience in both logical and emotional terms. Thus, the inner reflection on decision-making processes is important. If you feel uncomfortable, you will avoid making decisions and hesitate before repeating it. Reactions in chaos must be immediate as a rapid response.

The same goes for the collaborative part, where trust is one of the core ingredients. Trust is not only based on logic, but also on emotions. Research in psychology concludes that emotional trust is based on positive feelings and will decrease in situations with negative feelings. However, emotions do not influence trust if individuals are aware of the reasons for their feelings (Dunn & Schweitzer, 2005). One can say that this is common sense; however, it also emphasizes that conflict handling and collective reflections and dialogues on conflict or unpleasant situations are important. You will not get through life or any collaborative situation without conflicts or disagreements whether outspoken or tacit.

We still need to work on how we integrate emotional reflections into education, both individually and collectively. It might not be enough to get the students to reflect, which can be based on the logic of experiences rather than on the emotional sides.

What is important, however, is that there is focus on both the individual and the collaborative level. Collaborative efforts to support one another are likely to be necessary in potentially chaotic situations. Individuals carry the transformative elements from one context to another, collaborating to create a Gestalt or a common collective understanding. As individuals, we are in a constant process of forming and re-forming our interaction with the surrounding world, and therefore we need to focus on both levels in building resilience and empowerment among students.

While the reform universities mainly focused on the collective aspect of learning, the next wave should also include the individual growth and the self as identity development. We need to bring in values at a personal and collective level, and we have, in the PBL understanding, left out the individual/personal level. Collaboration consists among individuals and it is a mutual influence. The collaboration in the team will influence the individual, and the individual will influence the team.

11.6 Variation in Educational Practice

Another key message in this book is the variation in the learning process, the application of learning and teaching methods, and the curriculum structure (Ryberg & Kolmos, 2023). Variation in the educational practice combined with broader reflection modes could be a key to fostering knowledge, skills, competencies, and students' resilience. The educational system often becomes too uniform, and students do not learn to act and interact in different organizational settings or build up a more fundamental collaborative resilience if they are always introduced to the same learning methods.

Variations concern types of problems, including projects, learning outcomes, project duration, team size, the combination of projects and other teaching and learning methods, and interdisciplinary and transdisciplinary collaboration. Different types of problems include both academic and theoretical problems on one side and societal problems with identified stakeholders on the other. The level of difficulty is also relevant, from simple and complicated to complex and chaotic problems.

By learning in different ways, students also learn more about their own learning. Potentially they become further motivated by meeting people they are going to work with to solve the problem, or they are motivated by the abstract and theoretical thinking. Perhaps they feel that they control the learning best on their own or do the opposite in collaborative teams.

This is both about identifying one's own preferred learning approach, and also about experiencing and preparing for a variation of learning and organizational contexts. In work, there will be a variation of work organizations that sometimes work in parallel (Holgaard et al., 2025). This means that you should be flexible and can handle different project organizations within the same timeframe. This is something that we do not necessarily educate for, but this can be learned by a higher degree of purposeful variation in learning methodologies.

11.7 Academic Culture, Management and Push for Change

PBL represents certain sets of values, such as participatory processes where students influence and make choices; they become involved with social engagement, working on societal problems and in collaborative environments. To build such an environment also involves the academic and management cultures. The academic teachers must be curious together with the students, and facilitate and empower their curiosity.

Management plays an important role in supporting teachers, although this has not always been easy. During the past 15 years in Denmark, there has been a policy move from an elective management system to appointed systems. The power flow turned from a bottom-up flow between the 1970s and 2000s to a top-down flow today. This power structure and flow do not necessarily support a PBL culture.

Top-down-led educational development can lead to a very good and structured education system; however, those academics keeping things going on a daily basis can be forgotten. In order to run a PBL or CBL curriculum, engagement, visions, ideas, encouragement, energy, optimism, humor, and the courage to try out something new are all necessary. With a total top-down structure, I would claim that it will not be possible to implement a more student-centered curriculum. What history has shown is that it was the resistance stemming from the students' movement in the 1960s that drove the new pedagogy within the reform universities (see Chap. 3).

I am not convinced that the students of today will raise a critical voice. While some students in the US initially spoke out during the latest Palestine-Israel conflict, the Trump administration has since silenced these voices. Perhaps there will be a pushback from the students and the universities, and only time will tell. There is no doubt that a possible pushback will be involved with a new political agenda such as the 'woke' and 'Me Too' movements.

This is also part of democracy if it is combined with learning to step up and be able to solve conflicts in a personal and collaboratively responsible way. The MeToo movement has permeated workplace culture, even in situations that do not involve sexual harassment. There is a tendency to move problems up to a leader level to solve these issues instead of letting the involved individuals have a direct and responsible dialogue. In this way, the broader MeToo culture supports a top-down approach.

I would take it a step further and say that this contradicts a PBL culture, which needs to have an integrated and smooth conflict-solving approach in the collaboration, otherwise, the tasks and culture will not move ahead. Communication and dialogue are fundamental in creating a *Gestalt* of common understanding.

It is not easy to predict what will happen now with the young generation growing up. Other voices might push for a change; however, the original PBL and student-centered learning approaches are aligned with democratic management and dialogue in an academic culture.

11.8 And What about AI?

What is the impact of AI? This is one question I am sure the reader will have. The topic has been on my mind during the entire process of writing this book. Although it's outside my expertise, I have some reflections to share.

Higher education tries to embrace artificial intelligence. AI is here, and we have to deal with it in one way or another. I believe there are several aspects that we need to consider. First and foremost, students should possess the ability to stand independently without relying on AI. On the day that electricity goes out AI is of no use, so they need to be able to think and act independently.

The danger with AI is the fight for the truth, and AI is working on algorithms and the input of humans. There is already evidence that input data can cause bias in the conclusions drawn by the algorithms. Critical thinking will be one of the most important cognitive competencies in the future. There is a fight for the truth, and sustainability and climate change in particular are two hot political issues.

Having said that, AI should be allowed as part of the students' project teams, and they should learn how to interact with AI in higher education from a critical stance. AI can serve as another team member, providing input to the learning process, just as students engage in dialogue with their problems throughout the projects. This will be an advantage that could progress learning even further to obtain a perspective and discussion from an AI client that can participate in a critical dialogue within the teams.

Furthermore, in PBL there is the advantage of identifying problems and asking critical questions. The dialogue with an AI client will benefit from these questions that might lead even deeper into knowledge, and this does have to have a critical perspective.

An important concern is that when students reach the university level, we may encounter individuals who have not completed their assignments independently but instead have relied on AI and expert systems to succeed. The consequence of this is twofold. First, we receive students who are dependent on AI systems and cannot work on their own, and second, we need to work closer together with the school and high school systems to integrate a critical component in their use of AI. In principle, the universities should be ahead due to research, although the practice development that exists in the school and high school systems can sometimes work very efficiently.

More fundamentally, AI will change our whole society by increased the automatization of work. It will be challenging for universities to navigate in this landscape concerning both future engineering competencies and societal function.

11.9 Contribution to a New University Vision?

Visions embrace a holistic and comprehensive picture, including the role of universities in society, research, and the contribution to innovation and education. I have only addressed the educational part from a short historic introduction to perspectives on addressing chaos.

I do belong to a group that states that a university education should not only be about knowledge and competencies but is also about *Bildung*—the identity development and preparation for life. *Bildung* is the continental European concept of being educated in a broad understanding, as a responsible and moral citizen, and for being able to participate in society in a cultivated way. The identity aspect embraces the ability to partake in lifelong development as a citizen of emotional, moral, and responsible character. It is about the longsighted values of sustaining the world and not of a short-sighted profit, which dominates most cultures today. Barnet does not use the concept of *Bildung* but indicates the process of being, acting, and knowing that the 'being' is the formation of the identity (Barnett, 2010; Barnett & Coate, 2004; Barnett et al., 2001).

During life, our beliefs and values will develop as family and work continue to influence us. However, the fundamental thinking ought to be learned during education. We have our experiences and gather more as we continue to grow, but the truth is not our actual experiences. The truth is more a result of experience, dialogue, and social interaction combined with a more theoretical and factual data analysis.

We need to make a turnaround so that university education, along with the rest of the educational system, can become a driving force for this change. Developing technologies for war is a stupid and destructive mechanism, although I recognize that as long as war and other threats are out there, war technologies might be useful as a defense. As for the reform universities during the 1970s, again there is a need to learn democracy and value peoples' rights in a democratic society. This also involves a pedagogy of having the courage to stand up for this and trying out the application of various teaching and learning methodologies.

Examples of new learning practices exist, such as the Chaos Pilots (Kaos Piloter, 2026) and emerging institutions such as the Climate School (Erasmus University Rotterdam, 2024). Several university teachers integrate aspects of resilience in their courses by individual and collaborative reflection. There are minor experiments with transdisciplinary learning across the educational boundaries, which indicate an emerging trend, although they are hardly reported in the literature. And workshops and courses on handling disasters with stakeholders from civil administration, craftsman education and technical education start to emerge.

We must dare to speak about the difficult aspects, such as emotions, anxiety, and collaborative strategies for handling chaos. I hope that this book can give some inspiration and kick off such conversations and not at least develop new learning practices. This was my intention.

References

ABET. (2026). *Criteria for accrediting engineering programs, 2025–2026* https://www.abet.org/accreditation/accreditation-criteria/criteria-for-accrediting-engineering-programs-2025-2026/

Barnett, R. (2010). *Being a university* (1st ed.). Routledge.

Barnett, R., & Coate, K. (2004). *Engaging the curriculum in higher education* (1st ed.). Open University Press.

Barnett, R., Parry, G., & Coate, K. (2001). Conceptualising curriculum change. *Teaching in Higher Education, 6*(4), 435–449. https://doi.org/10.1080/13562510120078009

Dunn, J. R., & Schweitzer, M. E. (2005). Feeling and believing: The influence of emotion on trust. *Journal of Personality and Social Psychology, 88*(5), 736–748.

Erasmus University Rotterdam. (2024). *The Bildung climate school: The absolute example of education breaking boundaries.* https://www.eur.nl/en/news/bildung-climate-school-absolute-example-education-breaking-boundaries

Guntzburger, Y., Pauchant, T. C., & Tanguy, P. A. (2017). Ethical risk management education in engineering: A systematic review. *Science and Engineering Ethics, 23*(2), 323–350.

Habbal, F., Kolmos, A., Hadgraft, R. G., Holgaard, J. E., & Reda, K. (2024). *Reshaping engineering education: Addressing complex human challenges*. Springer Nature.

Holgaard, J. E., Kolmos, A., Friedrichsen, D., & Routhe, H. W. (2025). Interdisciplinarity, generic competencies, and traits in engineering practice from the perspective of early-career engineers. *European Journal of Engineering Education*, 1–17.

Kaos Piloter. (2026). *Kaos Piloter.* https://www.kaospilot.dk/kaospilot-education/

Kiehl, J. T. (2023). Climate chaos: A complex issue. *Psychological perspectives, 66*(1), 45–56.

Kurtz, C. F., & Snowden, D. J. (2003). The new dynamics of strategy: Sense-making in a complex and complicated world. *IBM Systems Journal, 42*(3), 462–483.

Meyer, T., & Reniers, G. (2022). *Engineering risk management.* Walter de Gruyter GmbH & Co KG.

Müller, J., Remmen, A., & Christensen, P. (1984). *Samfundets teknologi: teknologiens samfund.* Systime.

Pogue, D. (2021). *How to prepare for climate change: A practical guide to surviving the chaos.* Simon and Schuster.

Ryberg, T., & Kolmos, A. (2023). Hybridisation, variation and reflection: New PBL curriculum principles. In *PBL in a Digital Age* (pp. 27–38). Aalborg Universitetsforlag.

Servant, G. (2024). *Pedagogies of collapse: A hopeful education for the end of the world as we know it.* Bloomsbury Publishing.

Snowden, D. J., & Boone, M. E. (2007). A leader's framework for decision making. *Harvard Business Review, 85*(11), 68.

Verlie, B. (2019). Bearing worlds: Learning to live-with climate change. *Environmental Education Research, 25*(5), 751–766.

FSC
www.fsc.org
MIX
Papier | Fördert
gute Waldnutzung
FSC® C083411

Kolibri 360 GmbH
Ferdinand-Jühlke-Straße 7
99095 Erfurt, Deutschland
produktsicherheit@kolibri360.de